LITERATURE
AND
COGNITION

CSLI
Lecture Notes
Number 21

LITERATURE
AND
COGNITION

Jerry R. Hobbs

CSLI CENTER FOR THE STUDY
OF LANGUAGE
AND INFORMATION

CSLI was founded early in 1983 by researchers from Stanford University, SRI International, and Xerox PARC to further research and development of integrated theories of language, information, and computation. CSLI headquarters and the publication offices are located at the Stanford site.

CSLI/SRI International   **CSLI/Stanford**   **CSLI/Xerox PARC**
333 Ravenswood Avenue   Ventura Hall   3333 Coyote Hill Road
Menlo Park, CA 94025   Stanford, CA 94305   Palo Alto, CA 94304

Library of Congress Cataloging-in-Publication Data

Hobbs, Jerry R.
    Literature and cognition / Jerry R. Hobbs.
      p. cm. -- (CSLI lecture notes ; no. 21)
    Includes bibliographical references.
    ISBN 0-937073-53-9
    ISBN 0-937073-52-0 (pbk.)
    1. Discourse analysis, Literary. 2. Discourse analysis--Psychological
    aspects. 3. Literature--Psychology. I. Title. II. Series.
P302.5.H63 1990
808'.0014--dc20   90-1615
                                                                CIP

For William and Thomas

Contents

Introduction

Literature is first of all discourse. Therefore, it should be possible to apply to it methods and insights arising from the analysis of discourse. In the past two decades, considerable progress has been made in cognitive psychology and artificial intelligence in the study of discourse, particularly discourse comprehension. This work is motivated by the computer metaphor for mind. One hypothesizes that the mind is like a computer and asks what sort of computer it must be, given what it can do. This leads to an emphasis on possible mental representations of facts and beliefs and on possible computational processes that operate on these representations. From this perspective, every text, even the seemingly most ordinary, becomes a challenge. The simplest sentences raise deep problems. Since ordinary texts are so difficult, the field has remained largely immersed in them for all these years. But perhaps it is time now to step back a bit and ask if the perspective we have developed has anything to say about those extraordinary texts that constitute our literary tradition and about the issues that literary theorists have struggled with in trying to understand what it is to interpret such texts. This book consists of a series of my own attempts to address some of these issues.

I would like this book to be read both by cognitive scientists and by students of literature and literary theory. But because it is quite formal, it presents each group with complementary difficulties. Cognitive scientists, and especially researchers in artificial intelligence, may feel it is not formal enough, and, moreover, that it is only by avoiding adequate formality that one can even talk about literature, viewing literature as simply beyond present capabilities. In many instances, I outline some bit of world knowledge, describe a process in prose, and then assert that the process using the knowledge can find the desired interpretation of some passage. Those struggling with formal theories of commonsense reasoning and with computer implementations of language use will know that between such an account and a satisfactory formalism or computer program falls a very, very large and dark shadow, hiding more pitfalls than we can imagine. I can only say that I am aware of all of this, but I believe that the difficulties can eventually be worked out and that it is important to be able to take a long view from time to time to convince one's self that the path we are taking is the right one. A theory of discourse comprehension that can never hope to explain how we understand a sonnet is one that *must* be wrong, and even if we cannot work out the details now, it is reassuring if we can convince ourselves that we will perhaps someday be able to.

The problem that will confront the literary critic and literary theorist is that it is all *too* formal. Formal approaches always involve oversimplifications. In too many mathematical approaches to hard problems, the researchers at some point abandon their insights and start turning the crank, allowing themselves to be carried along by an elegant but oversimple theory of a very complex phenomenon. But description of any sort involves oversimplification. Whether we are constructing formal or informal descriptions, we must be careful that we have not eliminated from our account precisely those phenomena that are most interesting.

The literary scholar is likely to be especially suspicious because formal approaches in the recent past have eliminated from consideration some very interesting phenomena indeed. The New Criticism, for example, made impossible the consideration of the historicity of interpretation, and Structuralism seemed to reduce everything to a few binary oppositions. Is that happening here as well?

I believe it is not. To simulate even the simplest variety of discourse comprehension on the computer, it has been necessary to introduce representational formalisms such as predicate calculus that have a very great expressive power, to encode sometimes huge amounts of commonsense knowledge in these formalisms, and to devise very complex processes that operate on these representations and represented facts. At this point we are very far indeed from a small set of binary oppositions. Moreover, the very fact that much commonsense knowledge must be encoded forces us into an acknowledgement of the historicity of interpretation. Interpretation is impossible in the absence of a belief system, and in fact the primary focus of the theory of discourse interpretation presented here is the question of how these beliefs are used in interpretation, and hence, along with the text itself, determine the interpretations. But commonsense knowledge or belief systems necessarily vary from person to person, from culture to culture, and from era to era. The dependence of interpretation on a point in time is thus fundamental to what is presented here.

An important recent advance in literary criticism and literary theory is a heightened awareness of the need to make explicit, and problematic, the language and presuppositions behind the work and behind our analyses of the work. This point of view is in harmony with the perspective urged here. The methodology of discourse analysis presented in this book fairly *demands* that the language and presuppositions underlying a work be made explicit and consequently be examined. The methodology poses questions of a text that forces such an explicitation.

Another source of apprehension for the literary scholar may be that since we are viewing the mind as a computer, we are proposing a theory that is materialistic and mechanistic. Well, what can I say? It is. For those who find this disturbing, all I can do is urge you to rethink material and rethink machines. If the mind is a machine, that in no way means we need to value it less. The magnificent progress made by modern medicine has come from viewing the body as a machine, and that has certainly not led us to value our bodies less.

Another difficulty is the mere presence of all these mathematical formulas. Most literary scholars do not spend their professional lives reading such things, and in fact may have very little patience for them. For such readers, I have tried to write in a way that the formulas can simply be ignored. They are all followed immediately by a gloss in ordinary English. The glosses necessarily involve mathematical variables—the x's and y's many people stopped taking algebra to get away from. There's no helping that. But just refuse to be intimidated and think of them simply as proper names (although you'll see some rather strange entities getting proper names). If you glance up at the formulas as you read the glosses, you'll soon find reading the formulas themselves is pretty straightforward. Mostly, it's just English sentences in verb-subject-object order.

The problem of logical notation is especially severe in Chapter 4. You are urged, if you do give up on the formalism, rather than skipping to Chapter 5, examine Figure 4.1 and skip to Section 4.4. You should find the going easier there, and that is where the principal issues are discussed.

A final difficulty would be faced by the literary scholar who read the book and was completely convinced. He or she may begin to fear that literary criticism will become too scientific. Those with no mathematical interest or talent will be left out in the cold. Even if my wildest ambitions were achieved, there would be little danger of that. A formalism is a device that allows us to think and speak a bit more precisely than we otherwise might, and sometimes to reach conclusions that we otherwise might miss. That's all it is. Formality is a trick that can be learned. It can never replace insight. The person who has the insights will always occupy the central place in the study of literature, and of anything else.

The chapters of this book are largely independent and can be read in any order. There are two exceptions to this. It would be better to read Chapter 3 presenting the general framework of the theory of discourse interpretation before reading Chapters 4 or 5 on specific aspects of that theory, and it would be better to read Chapter 5 on the method of coherence analysis before reading the two examples applying the method in Chapters 6 and 7.

The first two chapters of the book lay out a theoretical framework, and in that framework address some fundamental problems concerning the nature of interpretation and the function of literature.

In the first chapter I present what I take to be the cognitive science view, at least for the purposes of investigation, of the structure of intelligent agents, and then use this view to respond to several positions that have been advanced by literary theorists in the recent past, including the New Critics, E. D. Hirsch, Stanley Fish, and Knapp and Michaels. I argue that much confusion in these discussions could be avoided if one were more explicit about the roles of both text and beliefs in interpretation.

The second chapter is a shorter piece, in which I rush too quickly through a number of deep issues concerning the possible functions of literature for creatures whose minds are computers. It might have been entitled more whimsically, "Will Robots Ever Have Literature?" The result of it all is quite conventional. The function of literature is (at least) just what Horace said it is—to delight and instruct.

The next three chapters provide more details about the theory of interpretation. The focus is on ordinary rather than literary discourse, but the processes described are required for arriving at interpretations of any sort of discourse, including literary, and the problems given special attention in Chapters 4 and 5—metaphor and coherence—are certainly of paramount interest in literary studies.

Chapter 3 presents the outline of the theory of discourse intepretation. The theory attempts to answer the question of how knowledge is used in the interpretation of discourse, and the chapter suggests a structure for such an inquiry.

Chapter 4 then applies this framework to the problem of interpreting metaphors. The three examples that are examined come not from literary works but from ordinary discourse, but the principles discussed apply as well to the interpretation of literary metaphors. The interpretation of metaphor is shown to be a matter of linking up predications from different domains in the right way, and then deriving the appropriate inferences. It is argued that both of these processes often occur simply as a by-product of the ordinary processes of discourse interpretation described in Chapter 3.

Chapter 5 focuses on the problem of how to characterize and recognize the coherence and structure of discourse. A number of coherence relations that can link adjacent segments of text are defined in terms of the inferences that can be drawn from the propositional content of the segments. Ultimately, these relations depend on causality, the figure-ground relation, and similarity. The coherence relations are then used to define recursively larger-scale structures of discourse, to explicate somewhat the intuitive notions of topic and genre, and to develop a method of textual analysis.

Chapters 3, 4, and 5 present a way of reading a text very closely. The next two chapters apply this way of reading to two literary works. Their discourse coherence and structure in particular are examined. The method of analysis will not necessarily lead to any brilliantly original interpretations. Rather it leads us to a deeper understanding of how the ordinary meaning of a literary work is achieved, both by the reader and by the writer. It makes problematic certain features of the text that might have otherwise gone unnoticed, and hence leads us to a finer-grained appreciation of the artist's mastery in creating the work's meaning.

Chapter 6 applies the methodology to a sonnet by John Milton. A sonnet has the advantage for us of being short enough that we can examine it with some thoroughness and of being tightly enough structured that it repays a very close reading. It is shown, among other things, how recognizing the coherence structure of the poem requires making assumptions (drawing implicatures) that are key to the work's meaning, and how Milton exploits certain local ambiguities in delineating the central tensions in the poem.

Chapter 7, coauthored with Patrizia Violi, attempts to tackle a much longer work, Gérard de Nerval's novella *Sylvie*, traditionally felt to be a very difficult text. Our analysis first considers the entire work on a very much less detailed level, but it shows that the method can yield insights about larger works as well, works in which the structure is not so apparent. We then microanalyze two sorts of selected passages, first, four key episodes of the underlying story, and second, some of the more confusing transitions between episodes occurring at different times. For both of these, we show how the central themes of the novella are reflected in the fine structure of the passages.

The motivating insight behind all of this work is that literature is a kind of discourse and that therefore theories of discourse interpretation ought to shed light on the reading of literature. The book closes with a short afterword that is based on a talk originally given at a panel, organized by Deborah Tannen, on "The Aesthetics of Conversation" at the Georgetown University Round Table on Languages and Linguistics in March 1982. In it I try to draw a parallel between what the best of literature and the best of conversation do for us, by looking at those cases where ordinary discourse is not so ordinary after all. It is argued that literature is a second-order effect on the already magnificent achievement of ordinary discourse, and that the best of literature, just as the best of conversation, is characterized primarily by the relationship that is created between the writer/speaker and the reader/listener.

Acknowledgments

I have profited in various aspects of this work from my discussions with Patrizia Violi, Michael Agar, Armar Archbold, Jon Barwise, William Chace, Melissa Holland, George Lakoff, Steven Mailloux, Geoffrey Nunberg, Helen Nussenbaum, Livia Polanyi, Paul Schacht, and the members of the Interpretation Seminar at Stanford University. None of these people is responsible for any errors that may be found in this book nor for the opinions expressed here. Some, in fact, would argue vigorously against most of what I have written.

The research described in Chapters 3, 4, and 5 was supported by grants from the National Science Foundation and the National Library of Medicine, and all of the research was supported by a gift from the System Development Foundation to the Center for the Study of Language and Information at Stanford University.

Chapter 1, "Against Confusion," was originally published in *Diacritics*, Vol. 18, No. 3, Fall 1988, pp. 78–92. Chapter 4 is a revision of a paper published in 1983 under the title "Metaphor Interpretation as Selective Inferencing: Cognitive Processes in Understanding Metaphor," in *Empirical Studies in the Arts*, Vol. 1, No. 1, pp. 17–34, and Vol. 1, No. 2, pp. 125–142. I am grateful to the editors of these journals for their permission to republish this material here.

1

Against Confusion

To an outsider, particularly to someone doing discourse analysis in an artificial intelligence (AI) framework, the recent controversies in literary theory concerning the nature of interpretation are quite puzzling. One camp claims that the interpretation of a text can be anything. The other side claims that there is a single correct interpretation. But all of this confusion can be swept away by a simple observation: in mathematical terminology, interpretation is a function of two arguments, the text and a set of beliefs.[1] In interpreting a text, one therefore presents not only an interpretation but also the set of beliefs that warrants the interpretation. One can then go on, if one wishes, to ask the separate question of whether one set of beliefs has a more privileged status than another. Viewed in this light, the controversies are as if one camp said that the mathematical operation of *multiplication* was hopelessly indeterminate because in the context of 2 the product of 2 is 4 whereas in the context of 5 the product of 2 is 10, while the other camp claimed that, no, the product of 2 is always 4.

---

[1]I will often use the terms "function," "argument," and "value" in their mathematical senses. In the expression $quotient(60, 12) = 5$, $quotient$ is the function, 60 and 12 are its two arguments, and 5 is its value for these two arguments. Whether I intend these meanings or the ordinary senses of these three words should be clear from the context.

AI provides us with a technical vocabulary that makes it possible to be somewhat more precise and detailed than is customary in discussing processes of interpretation. In Section 1.2, I present a framework, along with a corresponding technical vocabulary, that has proved useful in investigating discourse interpretation from an AI perspective. Among other things, it allows us to explicate the roles of intention and belief in interpretation. We will then be in a position to examine several characteristic views in literary theory in terms of the framework.

There has been a recent and widely discussed claim that it is incoherent to separate meaning and intention. Since this distinction is crucial in what I present, I begin in Section 1.1 by responding to this claim.

1.1   Meaning and Intention

Steven Knapp and Walter Benn Michaels (1982) have argued, or rather asserted, that meaning is an incoherent notion in the absence of an author's intention. It is certainly true that in the canonical case a text has an author who intends to convey something, and that something is what we call the meaning. "What did you mean?" and "What were you intending to say?" are often taken as equivalent. To reinforce this identification, Knapp and Michaels have us imagine that as we are walking along the beach, we see a wave wash up and, receding, reveal a poem written in the sand. We will believe either that the poem was written by some spirit of the sea capable of intentions, or that the marks in the sand resulted from some hugely improbable coincidence. Knapp and Michaels have the following intuition: "... in the second case—where the marks now seem to be accidents—will they still seem to be words? Clearly not. They will merely seem to *resemble* words" (p. 728). The marks have no author, are thus not language, and thus have no meaning.

This is the whole "argument." Unfortunately, I have the opposite intuition. It seems to me that the marks in the sand *are* words and *do* mean something. The event would not be remarkable otherwise. In any case, neither their intuition nor mine is worth very much, both being theory-laden. The example is so implausible it is doubtful whether anybody could have very firm intuitions about it. Let us consider three more commonplace examples to see if it is possible for texts to mean something independent of an author's intention. Here I will be appealing to the reader's everyday intuitions about the word "intention"; in Section 1.2, I work toward a more precise analysis of the term.

The first example is printer's errors. A favorite of mine appeared in a New York Times article on the voyage of the Pioneer 10 spacecraft beyond the solar system. Toward the end of the article, the writer intended to say, "Pioneer 10 carries a message ... in the form of a plaque designed to show ... the place and time where it began its long journey." Instead the newspaper printed, "Pioneer 10 carries a message ... in the form of a plague designed to show ... the place and time where it began its long journey." Let us suppose this was indeed a printer's error and not sabotage. The fact that what was printed does not correspond to any author's intention in no way diminishes our enjoyment of it, and it is hard to see how we could enjoy it if we did not first interpret it, that is, determine what it means. This means something, and it means something *other* than what the author of the article intended.

The next example comes from the world of computers. Before giving the example, I will give three negative examples for purposes of orientation. I log onto my terminal in the morning, and on the screen I see the text, "Good Morning!" It is a text and it has meaning, but I do not need to attribute intention to the computer or deny that an intention lies behind the text. The programmer, whoever and whenever, was the author, and the text means what he or she intended it to mean.

Next consider a computer program that generates random sequences of English words. We look over the output of the program and find some sequences that approach genuine poetry. There is too much distance between the program and the output for us to call the programmer the author. The words might have been read in from a file the programmer never looked at, and the random-number generator might have been a library subroutine whose code the programmer never inspected. But it is quite reasonable to say that the sequence of words is not a text at all, but simply an object on which we have chosen to impose some interpretation, as a kind of play, in much the same way as we might see the shape of a dog in a cloud. We certainly would not act on the content of the text. If we found the words "Ronald Reagan is a communist," for instance, we would not thereby come to believe that Ronald Reagan is a communist.

Next consider a program that "plans" its utterances, of the sort that has been implemented by AI researchers. It has a goal, that is, a logical formula or other data structure representing the condition to be achieved. It has knowledge, again in the form of logical formulas or other data structures, about what kinds of states or actions cause or enable what other kinds of states and actions. There is a process, called "planning," that uses this knowledge to decompose the goal into subgoals and these into further subgoals, until it derives a sequence of executable actions—in this case, the utterance. Again, there is too great a distance between the program and its output for us to call the programmer the author of the output. If the system is in practical use, say, telling us how to find something or how to use or fix an appliance, we had better take the utterance as a meaningful text and act on its meaning. But a reasonable case can be made (although many balk at this) that the program itself has intentions. If we want to be especially concrete, we can say the goals and subgoals are its intentions. The whole structure of the program is informed by the folk psychological theory and vocabulary of intentional action, making attribution of intention quite natural. In this case, the text has meaning, and it means what the program intended it to mean.

But let us now consider an example that falls in the middle of these three cases. We do not have to search far. A pocket calculator will do. Suppose I type in "1129.35 − 959.47", and the calculator responds with the text "169.88." I'm certainly not the author of the text; I might even be surprised at what I see. Neither the designer nor the manufacturer of the calculator could be called the author; the distance is too great; it is extremely improbable that either of them ever considered my particular subtraction problem. The sequence of numbers is a meaningful text; I interpret it using the same rules of interpretation I would use if a human had typed it out in response to my question. My interpreting it is not merely playful activity; I might enter it on my income tax return, sign my name at the bottom, and become legally responsible for my interpretation. Finally, we would not want to attribute intention to a pocket calculator. To do so would be a trivialization of the notion of intention and a consequent trivialization of the point Knapp and Michaels had hoped to make. The text "169.88" is a meaningful text with no human author and no intention behind it.

A final example that drives a wedge between meaning and author's intention is provided by Japanese linked poetry. In a group of three or four poets, one composes a stanza. Another poet makes up a second stanza related to the first in some way. A third poet composes a third stanza that is related somehow to the second but not necessarily to the first. The poets continue to alternate in this fashion for 36 stanzas, to produce a poem that goes through quite a number of twists and turns. It is quite common for a new stanza to force a reinterpretation of the preceding stanza, changing the implied locale, the circumstances, the gender and condition of the agent, and even the meanings of words. Very often, one suspects, the reinterpretation would surprise and delight the preceding stanza's author. A typical stanza thus has two meanings, one corresponding to its author's intention and determined by its link to the preceding stanza, and one constructed by the author of the following stanza and determined by its link to that stanza. Moreover, both meanings are essential to the working of the complete linked poem.

All three of these kinds of text are intentionless (or, in the third case, doubly intentioned), but they are hardly "accidental likenesses of language," and they have meaning. Though commonplace, they are admittedly marginal, but like many marginal phenomena they allow us to see clearly distinctions that are blurred or masked in more central cases. They show that meaning and author's intention do not coincide.[2]

---

[2]DuBois (1987) provides another example of meaning without intention—divination. A large set of texts is written by one person, with no detailed knowledge of the contexts in which they will be read. One of these texts is chosen by another person by some random means at a much later time and

There is another (uninvited) possible reading of Knapp and Michaels' article. They could be saying that to interpret something as a text, we must imagine an agent's intention as its source. The temptation of this position is clear. Since we are so adept at reasoning about human action, it often helps to imagine people in control where there aren't any.[3] But this is hardly necessary; the above examples show that we have ample experience with intentionless texts. So Knapp and Michaels, under this reading, could only be *stipulating* a new meaning for "intentional"; it is synonymous with "interpretable." Their argument then reduces to the following trivial one. We stipulate "x is intentional" to be equivalent to "x can be interpreted." Therefore, to be interpreted, an entity must be intentional.[4] The effect of this stipulation is to make the word "intentional" unavailable as a technical term; "interpretable" will suffice. But in the AI framework explicated below, "intention" and "interpretation" both turn out to be useful technical terms, and their meanings differ.

1.2 Interpretation

There is a technological aspect to AI—the effort to build smart computer programs—and a theoretical aspect. In the latter aspect, which is the one of interest in this book, one tries to discover general principles governing intelligent agents, regardless of how the agents happen to be embodied physically. This endeavor proceeds by means of a radical simplification. A computer program, or robot, or "cognitive agent," is constructed, or just imagined, to simulate, or duplicate, some intelligent behavior humans are capable of. This behavior is modeled in terms of formal symbol manipulation procedures. Questions about human capabilities, which are tangles of complex interactions and for which we have an inadequately precise vocabulary, are translated into questions

---

is interpreted with respect to some very particular circumstance.

[3]It is pleasant to speculate that this gratuitous attribution of human or humanlike agency is also the source of such phenomena as polytheism, hero worship, and conspiracy theories.

[4]Since, presumably, it is better to be wrong than trivial, I take it that the generous reading of Knapp and Michaels is my original one.

about the workings of the cognitive agents, for which we do have a precise, computational vocabulary and where we know, at least in principle, everything that is going on. This translation can isolate the core of an issue, suggest further lines of analysis, and frequently expose the falsity or tautologous character of an argument. We can often get crisp answers to mushy questions. Whether the crisp computational stories we tell about the cognitive agents project back to the human level is never certain. But if one is to argue that the crisp answers do not project back, one must say precisely how humans differ from the cognitive agents in a way that would make the projection fail. In any case, there is a long history in science of successful use of such idealizations.

The radical simplification is this. A cognitive agent possesses a set of beliefs. In AI this is generally called a "knowledge base" since one typically wants one's robot to believe true things. But because we will also want to include false and uncertain beliefs, opinions, values, heuristic strategies, and so on, we will call it a "belief system." One useful way of viewing a belief system is simply as a set of logical formulas encoding the agent's beliefs about the physical and social world in which it finds itself. The belief system includes not just general knowledge, but also a model of the immediate situation or environment—a theory of what is going on right now, including expectations, or beliefs about future events. The agent is linked to the world by means of various sensors and effectors. Its beliefs must be in accord with what it senses, and it will act in accord with its beliefs.

Next we can imagine a society of such agents, each with its own belief system. Suppose they can communicate, that is, produce and receive utterances via some medium. Then each agent's belief system must include beliefs as to what other agents in the environment believe and what beliefs it shares with them. Thus, beliefs must have more than just their content encoded; they also need to be tagged with information about who else believes them and, in particular, about what groups mutually believe them.[5]

---

[5] A set of people mutually believes a proposition if they all believe it, they all believe they all believe it, they all believe they all believe they all believe it, and so on, *ad infinitum* (Schiffer 1972).

Conventions may be represented in this fashion, including the conventions of language.

For purely computational reasons, we may assume that some particular subset of beliefs is active or in focus at any given moment. Only these beliefs are used by the agent's internal processes, although the agent also has means of moving beliefs into and out of focus. Alternatively, beliefs may have degrees of focus, where degree of focus determines the order of access to the beliefs by various processes.

The standard view in AI of the agent's procedure for generating utterances and other actions is that it is some sort of planning mechanism, as described above. The agent starts with a goal (an intention) and develops, or begins to develop, a plan of action, that is, a decomposition of the goal into subgoals, and these into further subgoals, ultimately yielding a sequence of actions, such as utterances, which it is believed will achieve the goal. As the actions are executed, the environment is monitored, and when unanticipated conditions arise, the plan is modified to accommodate them. Since utterances are typically intended to affect the beliefs of others, the planning mechanism, in designing the utterance, must take into account the beliefs of the other agents participating in the discourse, and especially those things that are mutually believed. Moreover, it must take into account the interpretation procedures that will be used by the other agents. What is presupposed by an utterance should be mutually known to the others or easily reconstructed by them. In particular, most of an utterance will depend on the conventional meanings of words and an implicit conventional theory about how utterances are understood. The less personal knowledge the participants have about one another, the greater the reliance that must be placed on conventions shared by a larger society to which they all belong.[6]

Thus, for the bare notion of intention, one substitutes a hierarchy of goals and a fairly complex planning and monitor-

---

[6]I should mention that all of this is independent of consciousness. High-level goals, like "Sell this used car," tend to be ones we would be conscious of; very low-level goals, like "Use the word 'reliable' here," tend to be ones we would not be conscious of. AI in general has little to say about the

ing mechanism, enabling a much more fine-grained analysis of what individual features of texts and other behavior are there to achieve.

AI work in discourse interpretation is characterized by a concern for specifying, with computational precision, how the listener makes use of his or her commonsense knowledge of the world and the immediate situation to interpret utterances, and in particular how utterances can be related to the speaker's presumed plan. Various accounts have been developed. In what follows I will, unsurprisingly, present my own.

We will assume the agent's interpretation procedure works by translating the utterances (the text), produced by another agent we will sometimes call the *author* and sometimes the *speaker*, into logical formulas and then drawing inferences from its belief system in such a way as to satisfy a set of requirements that specifies just what a "good" interpretation is.[7] What these requirements are is, as they say, a research question, but four very strong candidates are the following.

1. Utterances are anchored referentially in the mutual beliefs of speaker and listener, and reach out into the speaker's private beliefs in a bid to make new information mutually believed. This referential anchor must be identified and the new information must be recognized as such.

2. Words that are functionally related syntactically should be seen as congruent semantically. This constraint forces the interpretation of many instances of metaphor and metonymy. In the case of metonymy, an explicitly mentioned entity must be "coerced" into an implicit entity that satisfies the constraint. In the case of metaphor, certain inferences about an entity must be assumed or suspended to satisfy the constraint. In

America believes in democracy,

"believe" requires its subject to be a person, so "America" must be interpreted metonymically as standing for something like "the

---

experience of consciousness.

[7]In this assumption, we are taking positions on a number of controversial issues in AI and cognitive science, for example, the representability of knowledge in formal logic. These controversies, however, are not especially

people of America," or it must be interpreted metaphorically, acquiring for the occasion the relevant properties of persons. (See Chapter 4.)

3. Different segments of the text should be seen as coherently related, in a way that gives the whole text a unitary structure; this requirement for coherence in texts probably derives ultimately from principles of cognitive economy that people apply and that the agents should apply in attempting to make sense out of the world in general, principles involving things like causal linkage and assumptions that apparently distinct entities are identical. All of these constraints are sometimes violated, but where they are, the violation should be recognized; much of the delight that one derives from violations in literary works comes from our efforts to find a way in which the constraints are *not* in fact violated, to discover some hidden coherence. (See Chapter 5.)

4. The text needs to be related to the agent's theory of what is going on in the environment. Typically, but not always, this includes the agent's beliefs about the author's intention, or more generally, the author's plan as it unfolds in time; the agent should try to relate the text to what the agent believes the author is trying to accomplish.

This fourth point deserves expansion, since it is where interpretation and author's intention meet. The first thing to note is the phrase "what *the agent believes* the author is trying to accomplish." In the ideal case, the agent is entirely correct about the author's plan and cares about the utterance's relation to it. But like it or not, the agent, for all it knows, could be a brain in a vat, entirely deceived about what is going on around it. A real robot, especially during debugging, is often deceived in just this way, as its programmers manipulate its sensory inputs to test it. The agent can form good hypotheses about an author's intentions, just as it can about anything else in its environment. But it can never be certain about any of its hypotheses. The most

---

significant for the purposes of this chapter or the next. We could take other positions on these issues and construct a similar, though slightly different, framework and corresponding technical vocabulary to apply to the questions of interest in literary theory, and the results would be the same.

it could hope for is a consistent, parsimonious theory of the author's "psychological" life that will account for all the author's actions it perceives. So the author's intention plays at best an indirect role in the interpretation process: it plays a causal role in some observable actions, which the agent can then use, along with background knowledge, to form a belief about the author's intention. Only this belief can play a direct role in interpretation.

Moreover, among us humans there are many situations in which the author's or speaker's plan is of little interest to the reader or listener, and we would expect the same to be true for our cognitive agents. Someone in a group conversation may use a speaker's utterance solely as an excuse for a joke, or as a means of introducing a topic *he* or *she* wants to talk about. Very often two speakers in a discussion will try to understand each other's utterances in terms of their own frameworks, rather than attempt to acquire each other's framework. A medical patient, for example, may describe symptoms according to some narrative scheme, while the doctor tries to map the details into a diagnostic framework.[8] A spy learning a crucial technical detail from the offhand remark of a low-level technician doesn't care about the speaker's intention in making the utterance, but only about how the information fits into his own prior global picture. A historian examining a document often adopts a similar stance. In all these cases, the listener has his or her own set of interests, unrelated to the speaker's plan, and Requirement (4) involves no more than relating the utterance to those interests. In the conversations I have analyzed (see, for example, Hobbs and Evans (1980)), I have found this to be the case astonishingly often. Thus, not only is the role of the author's or speaker's intention indirect; it is frequently not very important.

The agent's interpretation procedure works by drawing inferences from its belief system, but two caveats are in order. First, inferences are drawn in a selective fashion, determined by what

---

[8]For example, I once took my young son to the emergency room for a cut hand. The doctor asked him what had happened. He said, "I went to Stevens Creek with my friend. His name is Paul." The doctor and I smiled at each other. "And there were some tin cans there." "Now we're getting somewhere," the doctor said.

will lead to a good interpretation. Instances of metaphor and irony are only the most obvious cases in which this control over inference is required. Second, the agent must often assume things to be mutually believed, for no other reason than that it will lead to a good interpretation of the text. David Lewis (1979) has called this process "accommodation," and we may call the proposition that is assumed an "implicature," since it is consistent with what H. P. Grice (1975) calls "conversational implicature."

We can summarize all of this in a single formula that is applicable beyond the details of this particular theory:

$$F(K, T) = I$$

An interpretation procedure F takes a knowledge base or belief system K and a text T, and produces an interpretation I. Each of these four elements requires some comment. In my comments, I will cease being fastidious about the distinctions between these agents and real people.

T: In general, there should be little dispute about T. Sometimes in conversation, one is not quite sure whether a nonverbal gesture is part of the text or just accidental, and in medieval manuscripts the words are often in doubt. But, for the most part, we can assume that the sequence of words that comprises the text is given.

Someone not familiar with recent literary theory might think this is all there is to say about T. But, as Stanley Fish has pointed out, interpretation goes all the way down. It is not a brute fact that a mark on paper is an instance of the letter "g," but is rather the result of interpretation. There have been, in fact, researchers in pattern recognition trying to make explicit the set of beliefs or interpretation rules that allows us to interpret an arrangement of lines and curves, or at an even lower level, an arrangement of pixels, as the letter "g."

Ultimately, in text interpretation as in every scientific or critical enterprise, we must bottom out in conventionally agreed-upon "evidence."[9] For text interpretation, this first involves a decision

---

[9]See Lakatos (1970). There is of course a significant problem concerning the epistemological status of "knowledge" acquired in this way, but because of their complexity, literary texts do not seem to be a good strategic locus

or an agreement *that* some physical object exists or *that* some physical phenomenon has occurred. This should not pose any problems. I doubt that any literary critic, as a critic, could seriously maintain that copies of *Ulysses* do not exist as physical objects, regardless of what one may take them to be. If we cannot accept the reality of trees, chairs, and books, it is hard to see why we should care about the feelings of Stephen Daedelus toward Leopold Bloom.

Next there has to be some conventionally agreed-upon account of how the physical entity presents itself to us. This does not seem problematic either, since one can express the account at as low a level as one pleases—for example, in terms of the impingement of light rays on the retina. Disciplines are defined by what they consider given and what they take to be problematic. Generally a literary critic will not be interested in interpretation processes below the level of the word or the letter. It would be acceptable to him or her to take as a fact that the first word of this sentence is "It." One can imagine circumstances, of course, in which it is crucial to determine whether a letter written in pencil is a "g" or a "q," and a microscopic examination may be required. Here the conventionally agreed-upon "facts" will be statements about the depth of the impression, the presence of bits of graphite, and so on.

Finally, one has to decide that this physical entity is to be interpreted as a text. This decision is part of a larger effort to construct the simplest theory, covering the most details, of all the entities one encounters; for some entities, the most economical theory is that they are texts. There are problematic cases, of course; an archaeologist has to decide whether scratches on a rock were carved by people or by a geological process. But the overwhelming majority of the things we decide to call texts give scant support for any alternative treatment. Once these assumptions are made, we are in the game defined by the above formula, and all of the following arguments apply.

Hence, we will assume that the text exists as a physical object, that there is a conventionally agreed-upon set of "facts"

---

for such an inquiry.

about what the object is at some level—whether pixels, letters, or words—and that a decision has been taken to regard it as a text and to apply interpretation rules to it. That is, we can take T to be given.

F: Some indications were offered above as to what the interpretation process looks like. AI researchers in discourse analysis have gone into greater detail in numerous articles, and more details are presented in Chapters 3, 4, and 5 of this book. It remains a big problem, but it is a healthy area of research. For the purposes of this chapter, we will assume the problem is solvable and ask what the consequences are. That is, we will assume F to be given.

It is important to note that there is a trade-off between F and K, between the interpretation process and the belief system used in interpreting. Any particular interpretive principle, such as "In Japanese poetry, the mention of cherry blossoms means that the season is spring," can be viewed as part of the interpretation process—as something we *do* when we interpret—or it can be viewed as a belief that is accessed by the interpretation process—as something we *use* when we interpret. There is no fact of the matter; we can choose either option. For the purposes of this chapter, we will choose the latter; interpretive principles are beliefs. Individual differences can also be accommodated in this way. It is quite possible that different people have different interpretation procedures, that they use radically different means to comprehend language. But even if this is true, then insofar as we are able to describe the interpretation procedures explicitly, we can factor out the differences, call them differences in belief, and let F be whatever is common to all interpretation procedures. Thus, F need not be indexed by *who* is doing the interpreting or *how* they choose to do it on a specific occasion. That is already encoded in K.

Finally, one might ask why F is a function in the mathematical sense of yielding only one value or result. Is it plausible to say that F applied to a single text and a single belief system will always yield a single interpretation? What about ambiguity? A purely formal way around this problem is to say that I can be not just a single interpretation, but a set of interpretations. But

I think a more satisfactory answer is possible. Generally, when we entertain different readings of an ambiguous text, we do so by shifting something in the belief system we are interpreting the text against. For example, when E. D. Hirsch (1967) sets out to convince his reader of the pantheistic interpretation of Wordsworth's poem, "A Slumber Did My Spirit Seal," he does so by spelling out Wordsworth's beliefs about "the immortal life of nature." In poems where we are given a few bare details that we can expand into a complete picture in several different ways, we can see our expansions as resulting from different implicatures, that is, different "beliefs" that we assume to be in K in order to accommodate the author.

I: An interpretation I is some formal representation of the content of the text that satisfies at least the four requirements for a good interpretation discussed above. It encodes the information conveyed by the text, the relevant inferences, and implicit structural relations that have been discovered among various elements. For most noncomputational purposes, a rough description in prose of the less obvious aspects will do.

There is, of course, more that one could say about a text than just what is contained in I. We can ask what someone would have to be like to produce such a text. We can ask what function the text performs in the larger social world. As I understand Hirsch (1967), these are questions about the "outer horizon" of the text. I is what I understand by his notion of "inner horizon."

K: The belief system K is intended to include the whole range of beliefs, from simple facts about the physical world to interpretive conventions for particular genres. Interpretation depends on context, and it is in K that the context is encoded. For different authors and different occasions, the agent will have different beliefs about the author's intentions, about what portions of the belief system are shared with the author, and about the current situation. In addition, on different occasions different beliefs will be in focus and different interpretations can result.

It has often been argued that context is unbounded, and that therefore it is impossible to formalize it.[10] Our knowledge is

---

[10] Mailloux (1985) has a recent and eloquent statement of this position.

certainly unbounded in the sense that indefinitely many propositions can be deduced from it, but this is hardly an argument against formalizability; deduction is well understood. The argument must therefore be that there are indefinitely many things one can say about a context beyond what can be deduced. It seems obvious to me, and I think most other AI researchers, that since we are finite creatures with finite access to our environments and an all too finite amount of time, there is only a finite amount of context that can be relevant to the interpretation of any situation. In fact, several large-scale efforts are under way to encode the knowledge an agent would need to understand everyday situations, and other projects are directed toward devising procedures for extracting from this knowledge just the parts that are relevant to any particular situation. The formalization of context is still an unsolved problem, but it is a vigorous area of research.

The belief system used in interpretation need not consist only of statements that are actually believed. A statement may also be embedded within a hypothetical context. This is required for understanding fictional texts and texts from other cultures and previous periods of our own, and also for understanding indirect proofs and other counterfactuals. The hypothetical statements enter into the interpretation procedure in exactly the same way as real beliefs, differing only in that they need not accord with what the agent perceives and in that the agent is less likely to act on them. We can flip among these hypothetical contexts with some facility, one time pretending we believe one thing, and another time something else. This is an important point for both discourse analysis and literary theory. Even though we often do not care about the speaker's beliefs in interpreting an utterance, at least as often we do care. In these cases, we can interpret the utterance not with respect to our own beliefs but with respect to our best guess of the speaker's beliefs. Insofar as we read literary works as a way of having conversations with the great minds of the past, it seems reasonable to interpret their texts with respect to *their own* belief systems, to the extent that we can surmise them. In brief, the beliefs used in interpretation do not have to be actually believed. We are not, as some writers try to cast us, prisoners of our own beliefs. We are prisoners of what we can imagine someone believing, and this gives us much more room for action.

Finally, there is no need to tie K to a real person. The belief system does not have to be *someone's* belief system. In this framework, it is merely the specification of a set of propositions. It can therefore be viewed as standing for the belief system of an ideal reader or an idealized author. It can be an author's real beliefs, or the beliefs he believes he shares with his audience, or the beliefs he wants his audience to think he has. It can be the set of beliefs a reader *should* bring to the text, whether or not anyone ever really does. It can be the set of beliefs that defines some "interpretive community." We can construct idealized, consensual belief systems against which to interpret texts of multiple or indistinct authorship, such as the Constitution or the Bible. By allowing such disembodied belief systems, we can abstract away from irrelevant vagaries of individual readers and writers.

Just what belief system should be used in interpreting a particular text depends on the purposes to which the text and its interpretation are to be put. In particular, what belief systems should be used in interpreting literary texts depends on the function of literary texts in our society. That issue is beyond the scope of this chapter, and largely beyond the scope of this book.

To summarize, then, we may assume that, in the equation, F and T are given and we must determine K and I. We have one equation in two unknowns. This of course does not determine either the belief system K or the interpretation I of the text, but it does place constraints on the possible K-I pairs. We cannot determine a belief system appropriate to the text simply by examining the text. We need to assume a particular interpretation of the text. Similarly, we cannot look at a text and determine its interpretation without making certain assumptions about the underlying belief system. When we understand or analyze discourse, we do so by hypothesizing a K-I pair. We assume an interpretation of the text and a portion of the underlying belief system that will support that interpretation. We can call this pair a "theory of the text." The equation expresses the fact that there are constraints on the possible K-I pairs, the possible theories of the text.

Consider an example. When I first read the opening line of Shakespeare's 68th sonnet,

Thus is his cheek the map of days outworn,

I had a very powerful image of an old man whose face was deeply wrinkled. These wrinkles were like the roads on the map of the life he had led. Later I read the footnotes. "Map" meant "symbol." "Days outworn" meant "ancient or classical times." The line meant that his face was the symbol of classical beauty— almost the precise opposite of my interpretation. I had interpreted the line against a belief system that included knowledge of Rand-McNally road maps and beliefs about the romanticization of old age. The function of footnotes is to tell the modern reader something of the belief system Shakespeare must have assumed he shared with his Elizabethan reader.

Another example comes from work that the anthropologist Michael Agar and I have done on some life history interviews of a heroin addict.[11] He is telling a story, and at one point he says,

Time was passing.
I was feeling worse all the time.

For most of us, there is no especially strong relation between these two utterances. But for the addict these sentences are elaborations on the same theme. If we are going to recognize this, we need to assume that very salient in his belief system is the fact that the passage of time implies that junk is running out and he is in need of another fix.

In specifying the details of K, different degrees of formality and precision are required for different purposes. At one extreme, about a decade ago I wrote a long and unreadable technical report (Hobbs 1976) giving an excruciating blow-by-blow account of what an interpretation procedure would do with one paragraph from *Newsweek*. The specification of the underlying knowledge base took 43 pages, and the account of what the interpretation procedure did with the text and the knowledge base ran to 58 pages. When one is talking not to computers but to people, as one does in discourse analysis and literary criticism, one can

---

[11]See, for example, Agar and Hobbs (1982).

focus on the difficult passages and state only the less obvious beliefs, as I did in the Shakespeare and the junkie examples.

There are many possible theories of a text within the constraints set by the equation. To decide among competing theories, or competing K-I pairs, we try to find the best K and the best I. I have already discussed some of the factors involved in determining how good an interpretation is. The junkie text provides an example. If we can discover the elaborative relation between "Time was passing" and "I was feeling worse all the time," the interpretation has greater structural coherence and is thus better than one that treats the two sentences as unrelated. There are various criteria that determine the appropriateness of a K. For literary texts one often wants the belief system that the author assumed he or she shared with the audience. Theorists who argue for the primacy of the author's intended meaning can be seen as arguing for the use of this belief system. One hypothesis about the belief system is then better than another to the extent that it generalizes over a larger number of texts by the same author or authors from the same culture. The Shakespeare example illustrates this point; the footnotes tell how Shakespeare and other Elizabethans used the words.

A text can be interpreted in many ways. Fish (1980) is adroit at showing how an initially outlandish interpretation can be made plausible, and this might be taken as an argument that a text can mean anything, or that an "interpretive community" can make a text mean anything. But this does not follow. To see how absurd this position is, let us consider what would be involved in constructing a "belief system"—in this case simply a lexicon—that would enable us to read *Paradise Lost* as *Hamlet*. "Of" would have to mean "who's." "Man's" would have to mean "there." (We'll ignore punctuation.) "First" would have to mean "nay." "Disobedience" would have to mean "answer." "And" would have to mean "me." "The" would have to mean "stand." "Fruit" would have to mean "and." But now we encounter a problem. "Of" would have to mean "unfold," but we've already said that "of" means "who's." We can get out of this by having context-dependent rules: following "fruit," "of" means "unfold." It is obvious that our difficulties become compounded the farther we go, and that as we approach the end, each rule for interpretation would be nearly as long as *Paradise Lost* itself. The point of this rather silly exercise is to demonstrate that the set of possible interpretations, large as it is, is really quite insignificant compared with the vast set of impossible interpretations. The requirement that a belief system must be constructed is really quite constraining, given the most rudimentary constraints on the content of the belief system. It means that the space of possible interpretations is one-dimensional rather than two-dimensional. We have one degree of freedom, but we do not have two. The difference is precisely the difference between having to stay on the highway and being able to drive all over the landscape.

It is important to emphasize that none of this unduly shackles the discourse analyst or literary critic. There is still plenty of room for his or her unique insights. As in any science, there are no constraints placed on the process of *arriving at* a theory. The constraints are applied in its *validation*. The analyst or critic can appeal to the full range of his or her knowledge of the author's culture and can use unconstrained ingenuity in constructing theories of a text. However, when it comes to validating a theory of a text or deciding among competing theories, he or she must convince us that the hypothesized belief system is appropriate and indeed supports the proposed interpretation. So for validity in interpretation, we do not need the author, as Hirsch (1967, 1976) argues; we only need to be explicit about the contributions of the belief system and of the text. All of this is not so different from standard practice. Even Fish, when he argues for the plausibility of an "Eskimo" reading for Faulkner's "A Rose for Emily" (Fish 1980, p. 346), does so by having us imagine that in Faulkner's belief system there is a belief that he is an Eskimo changeling.

Let us briefly examine several popular positions in literary theory in light of this framework. The New Criticism, and Wimsatt and Beardsley's position in particular, can be viewed as an attempt to standardize the belief system. The privileged belief system is an ideal one that includes only those beliefs or facts that an informed, but not too informed, reader would possess. It should include the conventions of language and presumably the facts about the world that are accessible to everyone, such as the fact that stones are not alive, but it should not include facts "about how or why the poet wrote the poem—to what lady, while sitting on what lawn, or at the death of what friend or brother." (Beardsley and Wimsatt 1954, p. 10). Of course, since there is such great divergence among various people's belief systems, one might ask whether the ideal is possible to achieve. For example, should it contain detailed knowledge of the *Odyssey*?

Generally, an author has a specific meaning to communicate to the audience. He or she has beliefs about what beliefs are shared with the audience, and so constructs the text upon this set of beliefs. Hirsch can be understood as saying that for literary texts the reader's task is to discover this belief system and to interpret the text with respect to it. There are many good arguments for granting this belief system a privileged status. An argument that is not good, however, is that only thus does a text acquire a determinate meaning. It already has a determinate meaning—determined by K and T both. Fix K any way you please, and the meaning is determined by T alone.

Knapp and Michaels, in their sequel to "Against Theory," "Against Theory 2: Hermeneutics and Deconstruction," characterize the hermeneutic position as one that posits a "verbal meaning" of a text which determines its identity but nevertheless allows it to be construed in various ways. Those adopting this position are seeking to explain how the same text can take on different meanings for different readers and different ages. Knapp and Michaels contend that it is arbitrary to choose verbal meaning as the criterion for textual identity, rather than, say, letters, or verbal meaning plus some bizarre additional rules of interpretation, and that the only coherent notion of meaning is the author's intended meaning. From the perspective of our framework, Knapp and Michaels are correct in saying that verbal meaning is an arbitrary choice—a text can be interpreted with respect to any K. The physical object, or rather the way it impinges upon our senses, is ultimately the only determinant of textual identity, and one can attempt to interpret it with respect to any K at all. The hermeneutic position is correct, or nearly correct, in that it isolates verbal meaning as the choice of K most appropriate for explaining the force of literary texts on readers through the ages. In effect, one partitions K into beliefs of interest and beliefs too low-level to be of interest. One interprets the physical object with respect to the latter set of beliefs and any two objects that yield the same interpretation— two copies of *Ulysses*, for instance—are for the purposes at hand viewed as identical. One then interprets this with respect to the beliefs of interest, including verbal meanings, or the conventional meanings of words. The beliefs of interest may coincide with the author's beliefs, in which case the interpretation will be what the author intended, or they may reflect the time and situation of the reader, in which case the interpretation may be quite different from anything the author ever imagined. In any case, Knapp and Michaels are simply wrong in saying that the author's intended meaning is the only coherent criterion for textual identity and the only coherent notion of meaning.

Fish, in the introduction to *Is There a Text in This Class?*, says, "In 1970 I was asking the question, 'Is the reader or the text the source of meaning?'" (p. 1). Within the framework we have developed, this is like asking of multiplication whether the multiplier or the multiplicand is the source of the product. The meaning or the interpretation I is a function of both the text T and the reader, parameterized as K. When Fish makes the provocative statement that there is no text until the reader writes it, he is really making the rather more mundane observation that there is more to K and less to T than one might have thought.[12]

The "facts" about the text are constructed, conventional facts, but that is not to say they are arbitrary. There are many "facts" that simply cannot be constructed. The "fact" that aspirin is a painkiller may be a constructed "fact," but it is not a possible constructed "fact" that LSD is a sleeping pill or that the Golden Gate Bridge collapsed in 1984. Our constructions, including our interpretations, are heavily constrained by the way

---

[12]He is also, of course, seriously underestimating the complexity of the *real* process of writing, something which is endemic in modern criticism, due perhaps in part to Eliot's (1920) false modesty in comparing the poet to a "catalyst."

the (not directly accessible) world is. There is no convention-free way to talk about the world, but that does not mean that there is nothing but convention. The world is still there to respond to our actions in ways beyond our control and to enforce a degree of mutual consistency with other agents. The world is experienced primarily (if not entirely) in the constraints it places on the interpretations we construct. The text exists as part of the world and is experienced as a set of constraints on what we can take the text to mean.

In 1979 Fish wrote that "meanings are the property neither of fixed and stable texts nor of free and independent readers, but of interpretive communities that are responsible both for the shape of a reader's activities and for the texts those activities produce" (Fish 1980, p. 322). This is an example, common in Fish's writings, of falsely posing several factors as mutually exclusive alternatives, rather than using the list of factors as a starting point in a detailed analysis aimed at discovering the contributions of each. It was stated above that a belief system contains not only the beliefs of the agent, but also an indication of who else holds those beliefs. For each fact P, it contains not just the fact P, but the fact $mutually\text{-}believe(S, P)$, where S is the set of people or agents among whom P is mutually believed. Fish's "interpretive community" is such an S. For an "interpretive community" S to be the source of an interpretation would be for the belief system upon which the interpretation is based to consist entirely of beliefs P for which $mutually\text{-}believe(S, P)$ is also believed. But it is obvious that there is seldom a single such S. Each reader belongs not to one but to a unique blend of many "interpretive communities." A variety of "interpretive communities," cultures, social organizations, shared and private experiences, and original ideas is responsible for a reader's belief system's being what it is, and thus they all contribute indirectly to the reader's interpretations. But it is only the belief system the reader uses that is directly responsible for the interpretations. By making the set of beliefs explicit, including the "interpretive community" associated with each of the beliefs, we can begin to tease out the contributions made by several "interpretive communities" to a single interpretation.

This chapter can be viewed as suggesting small but significant corrections to some views on interpretation that are commonly encountered in literary theory. The New Critics, Hirsch, and Fish all want to see meaning as a function of one argument. For the New Critics meaning depends on the text, for Hirsch on the author's intention. But neither of these computes. The text means nothing in the absence of rules to interpret it, and the author's intention is inaccessible until realized in some conventional way. By being explicit about the dependence of meaning on the rules of interpretation, or the conventions, one no longer has to argue about *which* rules or conventions determine *the* meaning of a text. The choice of a belief system to use is no longer an issue about "meaning" but an issue about the function of literature. Fish makes the opposite mistake. He discards the text and bases all on the reader or the interpretive community. Interpretations arise mysteriously, utterly unconstrained, out of interpreting activities. He supposes that interpretation can depend on only one thing, and recognizing its dependence on a system of beliefs, he is forced to banish what it is that is being interpreted. If we allow meaning to depend on two things, the text and a belief system, we are no longer forced into this implausible position.

2

Imagining, Fiction, and Narrative

The radical simplification of at least some branches of cognitive science is that instead of studying human beings in all their complexity, we look at cognitive agents (computer programs or robots) of which we have, at least in principle, a complete understanding. A cognitive agent is capable of certain perceptions and actions, and it is assumed to have goals and beliefs, which are encodings of logical expressions in a formal language. There are computational "inference" processes which operate on the logical expressions. Goals and beliefs are distinguished by the processes that operate on them; the processes act as though the beliefs were true and seek to find actions that will make the goals true. We as programmers, when we construct the cognitive agent, know the semantics of its formal language, and we link up the expressions with sensory and effector processes in the right way, given the semantics. After the agent has been embedded in a world for a while, it will acquire new beliefs, beyond what we have given it, and there will be a causal story, involving perception and inference, that will account for its "noninnate" beliefs. In our use of this idealization, we ask how much of the full complexity of human action we can construct out of such simple elements. Where we succeed, the result is not an account of how things actually are but only a proof of possibility.

This variety of cognitive science, proceeding in this manner, has made substantial progress toward an understanding of people's ordinary, everyday linguistic capabilities and activities. It has had less to say, however, about people's out of the ordinary, literary activities and achievements. In this chapter I would like to speculate a bit on whether the framework of cognitive science could lead to a better appreciation of the role of literature in human life. I will consider successively the possible functions that imagining, fiction, and narrative might have for a collection of communicating cognitive agents embedded in a world.

The imagination can be modeled as a set of logical expressions that are very much like beliefs in that they enter into the inferential processes in much the same way—hypotheses, for example, may be viewed as a kind of imagining—but with three crucial differences.

First, imaginings must be conscious, whereas beliefs may be unconscious. Cognitive science has little to say about the subjective experience of consciousness, but two features of consciousness can and should be modeled, the knowledge of one's own beliefs and "focus." In order to make inference processes computationally tractable, it helps to assume that some beliefs, including many recent perceptions, and some goals are in focus. Inference processes operate primarily or preferentially on the beliefs and goals in focus. In our radical simplification many properties of consciousness translate into properties of focus. Expressions that are imagined must then be in focus, whereas beliefs need not be. Walton (1990) disagrees with this, giving the example of a man who imagines his retirement consciously and unconsciously imagines that he is in good health when he retires. This is unconvincing, however. It is difficult to imagine a single proposition, just as it is difficult to believe a single proposition. Rather, we imagine and believe large complexes of propositions, and I would say that in his imaginings about his retirement, the man imagines in addition some properties that he himself would have, including the property of being intact.

The second difference between imagination and belief is that we cannot expect to tell the same kind of causal story for imagined propositions as for beliefs. Perceptions and inference certainly play a role in the origin of imaginings, but the tight connections required for belief need not be there, and in fact if they are, we are likely to call the proposition not imagined, but believed.

Third, the agent will not act as though imagined propositions are true. While the normal planning processes may be applied to imaginings just as they are applied to beliefs, the agent will not perform the indicated actions, or at least will not perform them in the expectation of achieving real goals.

There are at least two roles imagining plays in a person's life, that translate into corresponding possible roles in a cognitive agent's life.

1. We imagine things as a way of problem-solving by analogy, often as practice for or in order to work out solutions in leisure for situations that may arise in the future. The day before the Super Bowl at Stanford in 1984, the referees were out on the football field alone, pretending they were watching a play, and then pulling out the flag, trying to imagine every conceivable problem beforehand, so that during the Super Bowl their reactions would be immediate and reliable. The agent would similarly use time when no immediate action was required, to imagine or hypothesize problematic situations in order to work out the solutions beforehand and precompile them for rapid deployment should the situation arise in reality. Much play is of this nature. An agent that is intelligent enough to modify its environment will inevitably construct a world which, most of the time, is benevolent enough that the full capacities of the agent are not needed. At that point, the excess intelligence can be devoted to problems and activities that have no real consequences. That is, the agent will play. Often in play, we are working out the solutions in nonconsequential situations to simulated problems that we may sometime encounter in reality. This is a common observation about play.

2. Imaginings give us pleasure, make us angry, and evoke various other emotional reactions. Cognitive science has had little to say about the subjective experience of emotion. But we can talk about the combinations of beliefs and goals that are associated with various emotional states. Thus, pleasure is associated,

among other things, with a focused belief that one's goals will be fulfilled. The (very) radical simplification of emotion is then to identify the emotions with these goal- and belief-states. Under this view, the emotional reaction to imagining becomes very curious. The view suggests that belief is not crucial, that imagining is sufficient. Pleasure is associated with *any* focused proposition whose content is that goals will be fulfilled, whether the proposition is believed or just imagined. It is as though the emotional responses were not hooked up with goal- and belief-states quite right. It is possible that this function of imagination can be reduced to the first function, however. Insofar as the function of emotion is to impel us to generally appropriate actions without extensive reflection, often in situations in which there is no time to reflect, the emotional response to imagining can be seen as part of the analogical problem-solving process. We imagine a situation and perhaps practice a response, and the emotional reaction mediates between the imagining and the response, simply because that's the way it works in real situations.

A paraphrase of Horace's view of the function of literature provides a summary of all this: We imagine things to instruct and delight ourselves.

Let us now suppose we have a society of such cognitive agents. The society is constituted by conventions, or mutual beliefs, that arise from communication, agreements, and copresence, among other things. A mutual belief that P among a set of agents S occurs when each of the agents in S has a belief, that is, a logical expression of the form, say, *mutually-believe(S,P)*, together with the proper associated axioms for the predicate *mutually-believe*, allowing, for example, an agent to conclude individual belief from mutual belief. (If a society of agents discovered by communicating their experiences to each other that there were large areas of coincidence in their beliefs, thereby creating large areas of mutual belief, one can see that "truth" would be a useful concept for them to have.)

Mutual imagining, then, is like mutual belief except that it bottoms out in imagining rather than belief. That is, a set S of agents mutually imagines P when each of the agents in S imagine P, and they each believe that they all imagine P, and they each believe that they all believe that they all imagine P, and so on. The origin of any instance of mutual imagining will be either an explicit agreement or an implicit agreement by virtue of conventions in the society of agents. The functions of mutual imagining parallel the functions of imagining for the individual agent—cooperative problem-solving and "enjoying the pleasure of one another's company."

Mutual imagining raises the problem of how the rules of the game are to be communicated efficiently. How is it established exactly what is to be imagined? First of all there will be explicit provisions for the occasion. In one of Walton's examples, Jenifer says to Jason, "Let's pretend stumps are bears." Then there will be genre conventions. In certain games a long stick can always be a rifle; we needn't state that explicitly. But we cannot simply add these provisions to our belief systems, for that would likely result in inconsistency. For example, rifles have hollow barrels and sticks don't. What other changes need to be made to one's beliefs to carry on the imagining? A first guess would be that one makes the minimal change required to restore consistency. After all, the vast bulk of our knowledge is still appropriate; trees are still trees. This answer is of course unsatisfactory until a measure of minimality is defined reasonably precisely. Moreover, there may be several ways to reestablish consistency in one's beliefs that are of roughly equal measure. Consider the example of a cartoon: We learn that mice and ducks can talk, but dogs can't. What is the minimal change? One possibility is just that: mice and ducks can talk, and dogs can't. Another is that pets can't talk and other animals can. Another is that animals that walk on two legs can and animals on four legs can't. The rule we adopt will come into play when a bear comes on the scene. Can it talk or can't it? Even in solitary imagining the problem of what needs to be changed in the knowledge base arises. If a man imagines winning the lottery, he imagines the world to be otherwise the same. If he imagines having a harem, he has to make more substantial changes in his belief system.

Fictional discourse is an invitation to mutual imagining, in which the author provides explicit propositions to be imagined and the audience makes what they take to be the necessary minimal changes to the set of mutual beliefs the fiction is to be interpreted with respect to.

Most fictions are located in a tradition that sets the conventions about what is to be imagined and what is not. In realistic and romantic novels, for example, we are only to imagine those things that could be true for all we know. Thus, we can imagine that there was a person in Dublin called Leopold Bloom with all the described and narrated properties, but we would object if we were told that the British sovereign at the time was not Queen Victoria but King Victor. In science fiction, we can appeal to possible future technological progress to overcome inconvenient facts, such as the fact that habitable planets are vastly distant from each other being overcome by travel faster than the speed of light. Learning what these conventions are is part of what it is to become a full-fledged member of a culture, a part of what it is to come to have the right belief systems for the particular society of agents.

Certain works of fiction play games with the audience by challenging the conventions it expects to be operative. Fellini's movie "8 1/2" begins with the main character flying through the air. This event sets the viewer's expectations about what kinds of events can occur in this fictional world. Many bizarre things happen subsequently, but nothing quite this bizarre, and the viewer has no difficulty accepting the bizarre events. The reader of *Alice in Wonderland* soon learns that anything goes. Eggs and playing cards can talk, creatures can grow larger and smaller and can appear and disappear instantaneously. Probably the only way to read it is to view every rule in one's beliefs as subject to exception and treat every seemingly contradictory event as an exception. Another way of saying this: we ignore every real fact that proves inconvenient. Kafka's "Metamorphosis" forces the reader to carve a curiously shaped piece out of his knowledge base: A person can turn into an insect, but he retains his full human consciousness. Insects can be as large as people, but they still have trouble turning over when on their backs. And so on. From the initial events we would expect that anything goes, but in fact it doesn't. Much of the power of the story derives from the fact that for the most part the rest of the world remains the same, and how is such a creature to make its way in the world we know.

The functions of fiction are the same as the functions of mutual imaginings. Novels can be likened to experiments.[1] Situations that are more or less possible, but not actual, are set up and in a carefully controlled framework the author and the readers can explore the consequences of these situations.

Orthogonal to questions of fictionality are the central questions concerning narrative: What is narrative? And why, among the various forms of discourse, does narrative have its peculiar power over us? I believe the answers to these questions are related.

First, recall one more feature of our cognitive agents. They are planning mechanisms. They have goals, and they construct and execute plans to achieve these goals by decomposing the goals into subgoals and the subgoals into further subgoals until arriving at sequences or more complex arrangements of executable actions. Each of these decompositions of goals into subgoals derives from the agents' beliefs about what causes or enables what. That is, to achieve a goal G_1, an agent looks for some state G_2 that will cause G_1 and tries to achieve G_2. As it works through the actions in its plan, the agent monitors its environment to check on the success of its plan. When the plan fails, the agent modifies the subsequent steps in its plan to achieve its goals in another way and perhaps to repair the damage it has done.

A narrative is a species of discourse in which an entity, usually a person, is viewed as just such a planning mechanism, attempting to achieve some goal, generally in the face of some obstacle, and working out and working through the steps of a changing plan to achieve the goal. Since plans are constructed out of our beliefs of what causes and enables what, narrative presents a purported causal structure of a complex of events. It presents a character, like us a planning mechanism, maneuvering among these causal connections, attempting with or without success to

---

[1]This comparison was suggested to me by Jon Barwise (personal communication).

create a satisfactory outcome. This is perhaps the thrust behind that most trivial or most profound statement ever made about narrative, Aristotle's overquoted definition of the complete action required in tragedy as something which has a beginning, a middle, and an end. For Aristotle, what defined beginnings, middles and ends was causal necessity.

The peculiar power of narrative derives precisely from this. A narrative describes a planning mechanism planning its way toward a goal. We are planning mechanisms, continually planning our way toward goals. Thus, narrative presents us with situations and events precisely as we would experience them when we are most engaged with the world.

Much of what is most powerful in literature is a conjunction of the two categories—the fictional narrative. It is an author's invitation to the readers to a mutual imagining, to delight and instruct, by the creation of a possible world and possible characters striving toward goals, told in a way that directly reflects our own experience as we plan our way toward our goals in a world that denies us so much of what we desire.

3

A Theory of Discourse Interpretation

3.1 The Structure of the Theory

We understand discourse so well because we know so much. A theory of discourse interpretation must first and foremost be a theory of how knowledge is used in solving the interpretation problems posed by the discourse. This and other considerations suggest that the very large problem of discourse interpretation be carved into the six (still very large) pieces, or subtheories, listed below. Each subtheory is illustrated with an example relevant to one interpretation problem—the resolution of the definite noun phrase "the index" in the following text:

(1a) John took a book from the shelf.

(1b) He turned to the index.

3.1.1 Logical Notation, or Knowledge Representation

We must have a logical notation in which knowledge can be expressed and into which English texts can be translated. This problem has given rise to a large area of research, but I think the difficulties have been overstated. Typically, workers in this field have been trying not only to represent knowledge, but to do so in a way that satisfies certain stringent ontological scruples and canons of mathematical elegance, that lends itself in obvious ways to efficient computer implementation, and explains a number of recalcitrant syntactic facts as a by-product. If we decide to ignore these criteria or let some other part of the total system bear their weight, then most (though not all) of the problems of knowledge representation evaporate. (See Hobbs (1985).)

We will take first-order predicate calculus as our logical notation. It allows us to make and combine predications, and we can populate our logic with a rich set of predicates, such as *book* and *index*.

3.1.2   Syntax and Semantic Translation

Texts must be translated, sentence by sentence, into the logical notation. This also has been a major area of research for decades in linguistics and computational linguistics (Montague 1974, Woods 1970), and the solution has largely been worked out. The processes to be used are clear, the most commonly encountered syntactic constructions have been adequately analyzed, and current research is for the most part concerned with second-order refinements.

In our example, we may assume that syntax and semantic translation produce a logical form for sentence (1a) that includes the expression

(2) $book(b)$

and for sentence (1b) a logical form that includes

$index(i, z),$

where b, i, and z are existentially quantified variables. The processes of syntax and semantic translation can not be expected to determine what z is, that is, that i is the index of b. That is the work of other subtheories, described below.

3.1.3   Knowledge Encoding

The knowledge of the world and the language that is required to understand texts must be encoded in what may be called a "knowledge base." It will necessarily be huge, and the project of determining what needs to be represented, how to encode and organize it, and whether or to what extent it is consistent is correspondingly huge. Whether or not the project is tractable remains to be seen, but it is currently a healthy area of research (see Hobbs and Moore (1985); Hobbs et al. (1987); Lenat et al. (1986); Dahlgren (1985); Weld and de Kleer (1989)). We need not wait for the completion of this research before proceeding to a theory of discourse interpretation, for we can make general assumptions about how the knowledge is encoded, and we can assume specific (but not too specific) facts to be present in the knowledge base, as convenient. This is a way of isolating the problem of interest—how those facts are *used* by the interpretation processes.

We face a problem, however, in encoding this knowledge. It is difficult, if not impossible, to axiomatize in a consistent manner any domain more complex than set theory. Workers as early as Collins and Quillian (1971) noticed that it is a very powerful device to allow the following inconsistent set of axioms:

(3) $bird(x) \supset fly(x)$
        $ostrich(x) \supset bird(x)$
        $ostrich(x) \supset \neg fly(x)$

That is, birds fly, an ostrich is a bird, and ostriches don't fly. This is a much more economical representation than replacing the first of these axioms with something like

(4) $bird(x) \wedge \neg ostrich(x) \wedge \neg penguin(x) \wedge \neg kiwi(x) \wedge \neg emu(x)$
        $\wedge \ldots \wedge \neg injured(wing(x)) \wedge \neg dead(x) \wedge \neg newborn(x) \wedge \ldots$
        $\supset fly(x)$

That is, birds that aren't ostriches, penguins, kiwis, emus, injured, dead, newborn and so on, fly. The idea is that one can draw an inference as long as it does not result in an inconsistency, and that when an inconsistency does result, some means must be applied to decide among the inconsistent inferences. McDermott and Doyle (1980) developed a nonmonotonic logic in which the various exceptions of (4) are encoded with a special operator M, meaning "it is not inconsistent to assume that." Thus, (4) would be written

$$bird(x) \wedge M\,fly(x) \supset fly(x)$$

That is, if x is a bird and it is not inconsistent that x flies, then x flies. Nonmonotonic logic has since become a thriving area of research (Ginsberg 1987).

For the purposes of this book, however, it will be more convenient to keep the simple notation of (3) and complicate the calculus that manipulates it. For there are further reasons beyond the avoidance of inconsistency to be selective in the inferences one draws—there are too many true inferences that can be drawn in a specific situation and most of them are irrelevant. Consider the following text:

John couldn't find Mary's house.
He drove up one street and down another.

Among the inferences normally relevant to understanding this text are the facts that

Houses are visible objects.
Houses are located on streets.
People live in houses.

There are many more facts however that are not ordinarily relevant, and should not be "activated." For example:

Houses have roofs.
A house has a living room, a kitchen, several bedrooms, and one or more bathrooms.
Houses contain furniture.
Houses are made of such materials as wood, brick, stucco.
Termites sometimes attack wooden parts of houses.
Houses have exterior faucets to which hoses can be attached.
Houses tend to rise in value.

All of these things may be true of Mary's house, but the text (without further context) does not require them in any way.

A great deal of work in natural language processing can be viewed as addressing the problem of using the discourse itself to determine which inferences are relevant. The present proposal is that the relevant inferences are those required to solve various discourse problems, like recognizing the coherence structure of the text, forcing congruence between predicates and their arguments, and anaphora and ambiguity resolution.[1]

---

[1] Recent work by Sperber and Wilson (1986) presents a noncomputational attempt to characterize the relevance of utterances in discourse. The best

Text (1) provides a simple example. Suppose we hear only (1a). It is sometimes true that a book has an index, and sometimes it is relevant. But there is no reason, given (1a) alone, that we would necessarily want to draw the inference that John's book has an index. However, when we hear (1b), we can be sure that the inference is both true and relevant. Resolution of the definite noun phrase "the index" requires us to draw the inference that the book mentioned in (1) has an index. Note that we still need the normative knowledge that a book often has an index, even though the text mentions an index explicitly. If John had turned to the door, we would not have assumed the book had a door.

We may therefore assume that one of the facts in the knowledge base is the fact that books (at least sometimes) have indexes, and for simplicity write it as

(5) $(\forall x)\, book(x) \supset (\exists y)\, index(y, x)$

leaving it to the discourse operations to use this rule appropriately.

3.1.4   Deductive Mechanism

If we are to use knowledge stored as "axioms" in a logical notation, we must have some sort of theorem prover, or deductive mechanism, to manipulate these axioms and draw appropriate conclusions. This is not to say that language understanding *is*

---

interpretation of an utterance is then the one which gives it the greatest relevance. They take it that there is a set of contextually appropriate inferences associated with each interpretation of an utterance, and the best interpretation of the utterance is the one whose associated set of inferences is the largest and is derived with the least effort. The view expressed here and in Hobbs (1980) can be understood similarly. The discourse problems determine contextual appropriateness; an inference is contextually appropriate if it solves a discourse problem. One then selects that set of inferences which solves all the discourse problems most economically, that is, with the least effort. It's not clear what to make of Sperber and Wilson's proposal that the contextual implications should be *maximized*. I would contend that this is misleading, since one decidedly does not want to draw *all* the possible inferences. For example, the implication that Mary's house is probably rising in value should not make it a better interpretation to take "house" to mean a domicile rather than, say, a family line. But they could reply that the inferences we would not want to draw are not *contextually* appropriate, a notion they leave underspecified.

deduction, rather that it *uses* deduction. Deduction must be under the strict control of the discourse processes described below.

Automatic theorem proving is also a healthy area of research. There are many who have despaired of the possibility of devising efficient deductive procedures, but I think that despair is premature, for two reasons. First, parallel machine architecture, which the human brain surely possesses, is only now beginning to be understood. Second, we have little empirical data as to what classes of deductions are the most frequent in sophisticated language processing. It may be that special deductive techniques for the most common classes of inferences, together with parallelism, can overcome the efficiency difficulties.

One of the rules of inference the deductive mechanism will presumably provide is modus ponens. Thus, in our example, from

$$(\exists b)\ book(b) \hspace{3cm} \text{(There is a book } b.)$$

and

$$(\forall x)\ book(x) \supset (\exists y)\ index(y, x) \hspace{1cm} \text{(Books have indexes.)}$$

we will be able to conclude

(6) $(\exists b, y)\ book(b) \wedge index(y, b)$ (The book b has an index y.)

if the discourse operations require this inference.

Another feature is required of the inference mechanism, the ability to make assumptions. Very frequently, the best interpretation of a text cannot be completely supported by what is known, but could be if only a few assumptions were made. For example, to see the coherence in

John called Mary a Republican, and she insulted him too.

we must assume that the speaker believes that there is something wrong with being a Republican. We will call such an assumption an *implicature*. It is an accommodation the listener makes for the speaker, in order to maximize the coherence of the utterance. Such implicatures pervade interpretation. The simplest case occurs in the resolution of many pronoun references, where an identity must be assumed between two entities in order to maximize the coherence of a text (Hobbs 1979). More complex examples of implicature are given in the following chapters in this book. In Section 4.3.3 an implicature plays a key role in the interpretation of a complex, novel metaphor. In Chapter 6, various implicatures are central to the interpretation of a sonnet. In Chapter 7, several very large scale implicatures are required to see the coherence of a novella, and are crucial to its meaning.

3.1.5   Discourse Operations, or Specification of Possible Interpretations

A discourse presents us with certain "discourse problems," such as reference resolution, that must be solved if we are to be said to have understood the text. What counts as a solution can be specified in terms of inferences that can be drawn by the deductive mechanism from the propositional content of the sentence and the knowledge base. A possible interpretation of a sentence is taken to be a consistent combination of individual solutions to all of the sentence's discourse problems. The inferences that are relevant are then exactly those required by the "best" interpretation of the sentence, with "best" understood as explicated below.

We must therefore identify the discourse problems and, for each of them, specify what would count as a solution in terms of possible inferences. One discourse problem is the problem of discovering the referent of a definite noun phrase, such as "the index" in sentence (1b). A solution might be specified in approximately the following manner:

> The existence of an entity of the description given by the definite noun phrase can be inferred from the previous text and the knowledge base, and that entity is the referent of the definite noun phrase.

Thus, because the deductive mechanism using modus ponens, as in (6), can infer from the expression (2) in the representation of the previous text and from axiom (5) in the knowledge base that an index of book b exists, we assume that i is that index, thereby identifying z with b. The representation of text (1) now includes

$$(\exists i, b) \ldots \wedge \; book(b) \; \wedge \ldots \wedge \; index(i, b) \; \wedge \ldots$$

That is, there is a book b and an index i of that book.

Sections 3.2 and Chapters 4 and 5 will go into greater detail about specific discourse operations.

3.1.6   Specification of the Best Interpretation

The discourse operations only specify *possible* solutions to discourse problems, and there may be many. For example, in text (1) we may have the solution

> i is the index of book b,

or the solution

> i is the index of the first book listed in the bibliography of book b.

There must be principles that tell us that the first of these solutions is better than the second.

There has been very little work on this problem, although Nunberg (1978) has made a number of suggestions that deserve to be pursued. The basic idea is that we want to choose the most economical interpretation for the sentence or the text as a whole. Among the factors that count in determining economy are the complexity of the proofs supporting the solutions, the salience of the axioms used, and certain redundancy properties in the interpretation.

3.2   The Discourse Operations

Let us now look more closely at the discourse operations, closely enough to say just what the discourse problems are. In doing so, I would like to tell a story that suggests some logical necessity for just this set of discourse problems. We can divide the problems into those that arise in single sentences (whether or not they can be solved solely with information in the sentence) and those that involve the relation of the sentence to something in the surrounding context.

3.2.1   Within the Sentence

The logical form of a sentence consists of some logical combination of atomic predications, and an atomic predication consists of a predicate applied to one or more arguments. This suggests the following four classes of problems:

1. What does each argument refer to? This is the reference resolution problem; it includes the subproblems of resolving pronouns, definite noun phrases, and missing arguments. In ad-

dition, many problems of syntactic ambiguity can be translated into coreference problems (Hobbs 1982, Bear and Hobbs 1989).

2. Where the predicate is nonspecific, what predicate is really intended? An example of this problem is seen in compound nominals. What, for instance, is the implicit relation between the two nouns in "coin copier"? Other examples occur in denominal verbs and in uses of the possessive, the verb "have," and the prepositions "of" and "in."

3. How are the predicate and its arguments congruent? We may call the operation that seeks to answer this question *predicate interpretation*. In the simplest cases this just involves the satisfaction of selectional constraints, checking, for example, in "John believes in democracy," that John is a person as is required for the agent of "believe." When selectional constraints are not satisfied, there are two interpretive moves we can make. We can decide that the intended argument is not the explicit argument but something functionally related to it; this is *metonymy*. Or we can decide that the predicate does not mean what it ordinarily means, in the sense that some of the inferences one could ordinarily draw from its use are not appropriate in this instance; one example of this is *metaphor*. How metaphorical interpretations arise in interaction with other aspects of interpretation is the subject of Chapter 4.

4. Syntax tells us the *logical* relations among the atomic predications in the sentence, but frequently more information is conveyed. Consider for example the sentences

(7)   A car hit a jogger in Palo Alto last night.

(8)   A car hit a professor in Palo Alto last night.

Part of what is conveyed by sentence (7) is a causal relationship between the jogging and being hit by the car; inferring this relationship is essential to interpreting the sentence. We might call this the problem of determining *the internal coherence of the sentence*. Donnellan's (1966) referential–attributive distinction can be understood in these terms.

3.2.2   Beyond the Sentence

Next we can ask what the relation is between the sentence and the surrounding environment (the "world"). In more operational terms, what is the relation between the logical form of the sentence and some internal representation of the environment? This is of course such a huge problem it is certainly intractable. But there has been a great deal of work done in artificial intelligence on representing some aspects of the world as "plans" and attempting to specify how utterances relate to these plans. Such a plan may be a task model for some task the speaker and listener are executing jointly (Grosz 1977, Linde and Goguen 1978); it may be simply the speaker's presumed plan that led him to speak the utterance (Allen and Perrault 1980, Pollack 1986); it may be the listener's own conversational plan (Hobbs and Evans 1980); or it may be the plan of a character in a story that is being told (Bruce and Newman 1978, Wilensky 1983). We might call all of this the problem of determining the *global coherence* of the utterance.

One of the most important things that is going on in the environment is the discourse itself. It is important enough to be singled out for special attention. The listener, in interpreting the sentence, must determine, consciously or subconsciously, its relation to the surrounding discourse. We might call this the problem of determining the *local coherence* of the utterance. It is this problem that is the focus of Chapter 5.

4

Interpreting Metaphors

4.1 Metaphor Is Pervasive

I. A. Richards, in speaking of metaphor, said, "Literal language is rare outside the central parts of the sciences." (Richards 1936). But it is rare even in the central parts of the sciences. Consider for example the following text from computer science. It comes from an algorithm description in the first volume of Knuth's *Art of Computer Programming*, Vol. 1, p. 417, and is but one step removed from the domain's most formal mode of expression.

> Given a pointer P0, this algorithm sets the MARK field to 1 in NODE(P0) and in every other node which can be reached from NODE(P0) by a chain of ALINK and BLINK pointers in nodes with ATOM = MARK = 0. The algorithm uses three pointer variables, T, Q, and P, and modifies the links and control bits during its execution in such a way that all ATOM, ALINK, and BLINK fields are restored to their original settings after completion, although they may be changed temporarily.

In this text, the algorithm, or the processor that executes it, is apparently a purposive agent that can perform such actions as receiving pointers; setting, changing, and restoring fields; reaching nodes; using variables for some purpose; modifying links and bits; and executing and completing its task.

Nodes are apparently locations that can be linked and strung into paths by pointers and visited by the processor–agent.

Nodes also seem to be containers which can contain fields.

Fields are also containers which can contain pointers, among other things. In addition, fields are entities that can be placed at, or set to, locations on the number scale or in the structured collection of nodes.

Pointers, by their very name, suggest objects that can point to a location for the sake of some agent's information.

In fact, there is very little in the paragraph that does not rest on *some* spatial or agent metaphor. Moreover, these are not simple isolated metaphors; they are examples of large-scale "metaphor schemata," or "root metaphors" (Lakoff and Johnson 1980), which we use to encode and organize our knowledge about the objects of computer science. They are so deeply engrained that their metaphorical character generally escapes our notice.[1]

The pervasiveness of metaphor was noted as early as the eighteenth century by Giambattista Vico (1744 [1968]) and by Jeremy Bentham (cf. Ogden 1932). In our century, this observation has been the basis for a rejection of Aristotle's and Quintillian's views that metaphor is mere ornament, and an elevation of metaphor to an "omnipresent principle of language" (Richards 1936) and "the law of its life" (Langer 1942). Richards argued that metaphor involved complex interactions between two domains, which he called the "tenor," that which is being described, and the "vehicle," that which it is being described in terms of. The tenor is seen in a perspective provided by the vehicle, either bringing to the fore certain aspects of the tenor or allowing the tenor to be viewed in ways that would not have been possible without the metaphor.

---

[1] I have occasionally had a computer scientist argue that some of the metaphors, e.g., the "variable as container" metaphor, were not metaphors at all but true descriptions of physical reality. To see that this is not the case, note that when we place a value in a variable, its previous value is no longer there; we did not have to remove it. (I once had a beginning FORTRAN student who was puzzled by this very fact. He had not yet learned the limits of the metaphor.)

As we saw in our example, spatial metaphor especially is pervasive. Jespersen (1922) remarked on this. For Whorf (1939 [1956]) it was a key element in his view that language determines thought: the spatial metaphors provided by one's language determine how one will normally conceptualize abstract domains. Urban (1939) saw in the use of originally spatial words for more abstract concepts an "upward movement" of language from the physical to the spiritual. More recently, Clark (1973) examined the physical and psychological motivations behind our most common spatial metaphors for time. In Hobbs (1976) there is an attempt to exploit the pervasiveness of metaphor in a computational framework; the present chapter continues the attempt. In Jackendoff (1976) we find a similar effort in theoretical linguistics. The most extensive recent treatment of metaphor in everyday language is found in Lakoff and Johnson (1980); they identify the root metaphors that underlie our thinking about a vast array of domains, and argue that we can understand the domains only by means of these metaphors. The fundamental insight that informs all this work is this: metaphor is pervasive in everyday discourse and is essential in our conceptualizations of abstract domains.

In this chapter I wish to explore how metaphors and metaphor schemas might be treated in a computational setting, from the perspective of artificial intelligence, in a way that accommodates the fundamental insight. In Section 4.2, certain interesting previous proposals concerning metaphor are examined within the framework outlined in Chapter 3. In Section 4.3, three successively more difficult examples of metaphors are considered—first a simple metaphor, next a metaphor schema that has become a part of the language, and finally a novel metaphor. The aim is to discover some of what is needed to represent and reason about metaphorical usage. In Section 4.4, a number of issues of classical interest are examined in light of this approach.

4.2   Some Previous Approaches

In Chapter 3 the following model of language processing was described: A text is translated by a syntactic front-end into pred-

icate calculus formulas, and those inferences are drawn that are
necessary for solving the discourse problems posed by the text.
The inference process is selective and driven by a collection of
discourse operations that try to do such things as resolve pronoun and definite noun phrase references, find the specific interpretations of general predicates in context ("predicate interpretation"), reconstruct the implicit relation between the nouns in
compound nominals, and recognize coherence relations between
adjacent segments of the text. The operations select inferences
from a large collection of axioms representing knowledge of the
world and the language. Associated with the potential inferences
are measures of salience which change as the context changes.
These help determine which inferences are drawn by the operations and hence how the text is interpreted. The control structure is such that the system does not try to solve the discourse
problems independently, but rather seeks the most economical
interpretation of the sentence as a whole.

It is often advanced as an argument against a particular formal approach that it does not take context into consideration.
As Black (1979) has emphasized, metaphors occur in some context and must be interpreted in that context. It does not make
sense to ask about the interpretation of a metaphor outside of
a context. That is not an argument against the approach used
here. On the contrary, the framework outlined above is specifi-
cally designed to formalize a notion of context, and to provide a
way of interpreting expressions in context.

A number of previously proposed approaches to metaphor
interpretation can be viewed from the perspective of this framework as a matter of selecting the appropriate inferences, although
none of them had adequate means for dealing with the context
dependence of the selection process.

In *The Art of Rhetoric* (III.II.12), Aristotle said that "clever
enigmas furnish good metaphors; for metaphor is a kind of enig-
ma." In a sense, then, the idea of metaphor interpretation as
problem solving—like most other ideas—is originally due to Aristotle.

More recently, in computational linguistics, the earliest detailed proposal for handling metaphor was that of Russell (1976).

Her proposal concerns abstract uses of verbs of motion and involves lifting selectional constraints on the arguments of the verb while keeping fixed the topological properties of the motion, such as source, path and goal. Thus, to handle "the ship plowed through the sea," one lifts the restriction on "plow" that the medium be earth and keeps the property that the motion is in a substantially straight line through some medium. Russell exemplifies an approach that finds its most complete development in the work of Levin (1977), but it is also seen in linguistics in the work of Matthews (1971) and Kahn (1975). Metaphor is treated as a species of semantic deviance; selectional constraints are lifted until the expression can plow through the interpreter without difficulty. One can view a selectional constraint as a particular kind of inference. Thus,

(1) $plow\text{-}through(x, y) \supset earth(y)$

That is, if x plows through y, then y is earth. Then lifting this constraint is equivalent to not using (1) to draw an inference about the substance that is being plowed.

But the problem of interpreting "the ship plowed through the sea" is not just to avoid rejecting the sentence because the sea is not earth, but to notice the similarity of the wedge-shaped plow and the wedge-shaped bow of a ship and the wake that each leaves, and perhaps more importantly, to take note of the ship's steady, inexorable progress. In short, metaphor interpretation is less a matter of avoiding certain inferences than it is a matter of selecting certain others. Any approach to metaphor that does only the first of these is not a way of interpreting metaphors, only of ignoring them. Under this view, the fundamental insight about metaphor is simply bizarre and inexplicable.[2]

Several more recent approaches can be seen as aiming toward the selection of an appropriate set of inferences. For Miller (1979), the basic pattern of metaphor is given by the formula

(2) $G(x) \supset (\exists F)(\exists y)(SIM(F(x), G(y)))$

In words, this means the following. A predicate G is applied metaphorically to an entity x. To interpret the metaphor, one

---

[2]For further arguments against this approach to metaphor, see Nunberg (1978).

must discover a property F which literally describes x, an entity y which G literally describes, and the similarity between $F(x)$ and $G(y)$. By similarity, Miller means that there are "features" which $F(x)$ and $G(y)$ share.[3] The notion of "feature" is subsumed by the AI notion of "inference." Thus for Miller interpreting a metaphor $G(x)$ is a matter of selecting the inferences that one can draw from G that can also be drawn from the known (literal) properties of x.

There have been a number of recent proposals which may be viewed as specifications, prior to interpretation, of which inferences are the best to select. One proposal is that of Ortony, who also uses the notion of "feature." Ortony (1979) has suggested a breakdown of the knowledge about the vehicle and the tenor into classification facts, other high-salience facts, and low-salience facts. Classification facts are not transferred from the vehicle to the tenor. Thus, from "John is an elephant" we do not infer that John is a (nonhuman) animal. What get transferred from the vehicle to the tenor are other high-salience facts whose correlates in the tenor are of low salience. It is a high-salience fact that elephants are large, whereas John's size is generally of low salience. The effect of the metaphor is to bring to the fore this low-salience fact about John. That is, one draws the high-salience inferences associated with the vehicle that are not contradicted or confirmed by high-salience inferences about the tenor.

Carbonell (1982), working in an artificial intelligence framework, proposes pre-packaging the inferences associated with Lakoff and Johnson's root metaphors, recognizing on the basis of explicit content which "package" or root metaphor is being tapped into, and then drawing all the inferences in the package that are not explicitly contradicted by the text.

In view of the close relationship that is generally asserted to exist between metaphor and analogy, the work in artificial intelligence that should be most relevant to a study of metaphor is research on analogical reasoning. There are a number

---

[3]This is a weaker notion of similarity than Tversky's (1977) which also takes into account features that are not shared.

of examples. Evans (1968) wrote a program for solving geometric analogy problems. Kling (1971) built a system for proving theorems in ring theory by examining proofs of analogous theorems in group theory (a class of analogies that forms the basis of Galois theory (cf. Artin 1959)). Most of this work either has been conducted at too specific a level to be of use in our work on metaphor, or where the specific domain has been abstracted away from, has been too general to offer any new insights.

An exception to this is the work of Winston (1978). He presents an algorithm in which properties are transferred from the vehicle to the tenor if they are extremes on some scale, are known to be important, or serve to distinguish the vehicle from other members of its class. Thus, properties of elephants that are not shared by other animals would be transferred. Again, one can view the transfer of a property from the vehicle to the tenor as an inference one selects, and what Winston has suggested are criteria for selecting these inferences.

Gentner (1983) presents evidence that relations are more likely than attributes to be transferred from the vehicle to the tenor. That is, inferences are more likely to be selected if they involve a two-place predicate rather than a one-place predicate in the consequent. Thus, from the simile "the atom is like a solar system" one is more likely to infer that electrons go around the nucleus (a two-place predication) than that the nucleus is yellow (or roughly spherical).[4]

Toward the end of the paper cited above, Carbonell (1982) suggests a more refined classification of possible inferences. Inferences about goals and plans of agents and causal facts are most likely to be transferred from the vehicle to the tenor. Somewhat less likely are functional attributes, temporal orderings, and structural relations, and least likely, almost never relevant, are physical descriptive properties and object identity. It is not surprising that this should be the case, since the function of metaphor is usually to make sense of some abstract domain.

---

[4]There has been other work on metaphor by psychologists. A good review can be found in Ortony, Reynolds and Arter (1978).

All of this research seeks to specify certain classes of inferences that are typically transferred—on the basis of salience, arity of the predicates, convention, semantic content of the inferences, and so on. But these approaches suffer from the fact that they do not explain how context influences the interpretation of metaphors. None takes into account the text in which the metaphor is embedded.[5]

The approach taken in this chapter is to subsume the metaphor interpretation problem under the more general problem of making sense of a discourse as a whole. The discourse operations a natural language processor must possess anyway—operations like the recognition of local coherence, predicate interpretation, and compound nominal interpretation—will often serve to pick out the relevant inferences in cases of metaphor. Often the correct interpretation of the metaphor will simply "fall out" as a by-product of other interpretation processes.

Before the examples are presented, it should be pointed out that metaphors operate primarily at the conceptual level, and we will be dealing at all times at the conceptual level, not at the surface linguistic level. At the conceptual level, we talk about "predicates," not "words." Although we will generally have, for every word, a predicate of the same name, the predicate should not be thought of as exhausting what is conveyed and suggested by the the word. Rather, we should think of the word as corresponding to the possible sets of inferences that might be drawn because the word has been used in a particular context. That is, words do not merely translate into a single expression in a formal notation; they trigger an inference process that could result in any one of a large set of possible expansions in this notation. Hence, we have not stripped words of their mysterious quality, but rather translated the mystery into the mystery of choosing the right set of inferences.

---

[5]If we imagine salience as something which varies with context, then Ortony's proposal can be viewed as depending on context, but it is a rather blunt sort of dependence. Carbonell's choice of the pre-packaged root metaphor is dependent on explicit context, so this step in his algorithm at least is context-dependent.

4.3   Three Examples

4.3.1   A Simple Metaphor

Let us now consider how a simple metaphor would be interpreted in our framework.

(3)   John is an elephant.

Let us suppose our initial logical representation for this is

$elephant(J)$

There are a number of things we might infer from the fact that some entity is an elephant. Among the axioms allowing such inferences would be

$(\forall x)\ elephant(x) \supset large(x)$

$(\forall x)\ elephant(x) \supset has\text{-}trunk(x)$

$(\forall x)\ elephant(x) \supset good\text{-}memory(x)$

$(\forall x)\ elephant(x) \supset thick\text{-}skinned(x)$

$(\forall x)\ elephant(x) \supset clumsy(x)$

That is, an elephant is large, has a trunk and a good memory, and is thick-skinned and clumsy. The problem we are faced with in interpreting (3) is the problem we are always faced with in interpreting a text—determining which inferences it is appropriate to draw from what we've been told. Depending on the situation, we may want to infer $large(J)$ or $good\text{-}memory(J)$. The inference that John has a trunk is presumably rejected because of strong reasons to believe the contrary.

Which inferences *are* appropriate will depend on context. Example (3) contains insufficient context to allow precise interpretation. But we can embed it in a text in which discourse operations become decisive. For example, in

(4)   Mary is graceful, but John is an elephant.

coherence considerations force the interpretation. In order to recognize the contrast coherence relation (see Chapter 5) indicated by "but," we must draw the inferences that John is clumsy, and thus not graceful. Other possible inferences about elephants are not drawn, not so much because they would result in an inconsistency, but because no discourse problem requires them to be drawn. Other texts would force other inferences. Consider

Patricia is small, but James is an elephant.

Susan forgets everything, but Paul is an elephant.

Jenifer is subtle, but Roger is an elephant.

The inferences associated with the explicit predication in the metaphor (4) are of three classes. There are those inferences that are definitely intended—for example, the inference $clumsy(J)$ from (4). These "ground," or establish a firm basis, for the metaphor; they are what warrant it. Then there are those inferences that are definitely not intended and are inappropriate to draw, the disparities, such as $has\text{-}trunk(J)$. Finally, there are inferences that lie in-between, such as $large(J)$, which may or may not be intended by the speaker and may or may not occur to the listener. Much of the power of a metaphor derives from this third class of inferences—the other things that are suggested by the metaphor beyond its ground or firm basis. In fact, even the inappropriate inferences of the second class lend power to the metaphor, since the very denial of something suggests its possibility. The calling up and rejection of the image of a elephant in interpreting (4) may leave its trace.

4.3.2   A Spatial Metaphor Schema

Metaphors that tap into our spatial knowledge are especially powerful since our knowledge of spatial relationships is so extensive, so rich, and so heavily used. As soon as the basis for the spatial metaphor is established, then in our thinking about a new domain we can begin to borrow the extensive machinery we have for reasoning about spatial relationships. For example, once I say that

(5)   The variable N is at zero,

and interpret it as

(6)   The value of the variable N is equal to zero,

then I have tapped into a large network of other possible uses. I can now say

N goes from 1 to 100

to mean

The value of N successively equals integers from 1 to 100. I can say

N approaches 100

to mean

The difference between 100 and the value of N becomes smaller.

N can now *stay at* a number, *move from* one number *to* another *through* several others, be *between* two numbers, be *here*, be *there*. Variables can be *scattered along* an interval, they can *follow* one another *along* the number scale, they can be *switched*. In short, by means of the simple identification of (5) and (6) we have bought into the whole complex of spatial terminology.

In terms of our framework, what we mean when we say that our spatial terminology is an intricate network is that there are a great many axioms that relate the various spatial predicates. The concept of location—the predicate *at*—is at the heart of this network because so many of the axioms refer to it. For example, associated with the predicate *go* we might have an axiom like

$$go(x, y, z) \land at'(w_1, x, y) \land at'(w_2, x, z) \supset change(w_1, w_2)$$

That is, if x goes from y to z and w_1 is the condition of x being at y and w_2 is the condition of x being at z, then there is a change of state from w_1 to w_2. Similarly, part of the meaning of "switch" could be encoded in the axiom

$$switch(x, y_1, y_2) \land at'(w_{11}, y_1, z_1) \land at'(w_{12}, y_1, z_2)$$
$$\land at'(w_{21}, y_2, z_1) \land at'(w_{22}, y_2, z_2)$$
$$\supset change(w_{11}, w_{12}) \land change(w_{22}, w_{21})$$

That is, if x switches y_1 and y_2 and w_{ij} is the condition of y_i being at z_j, then there is a change from condition w_{11} to condition w_{12} and a change from condition w_{22} to condition w_{21}.

We were able to establish the metaphor "a variable as an entity at a location" simply by identifying (5) and (6). In our formalism we can establish the metaphor with similar simplicity by encoding the following axiom:

(7) $variable(x) \land value'(w, y, x) \supset at'(w, x, y)$

That is, if x is a variable and w is the condition of y being its value, then w is also the condition of x being at y.

Axiom (7), identifying "is the value of" with "is at," gives us entry into an entire metaphor schema and enables us to transfer to one domain the structure of another, more thoroughly understood domain.

The discourse operation of *predicate interpretation* uses axioms like (7) to arrive at interpretations of certain metaphorical expressions. The idea behind it is that most utterances make very *general* or ambiguous sorts of predications and that part of the job of comprehension is to determine the very *specific* or unambiguous meaning that was intended. Thus, someone might make the general statement

I went to London,

expecting us to be able to interpret it as

I flew to London in an airplane,

rather than interpreting the going as swimming, sailing, walking, or any of the myriad other manners of going. In the case of (5), we are expected to determine which of the many ways one thing can be *at* another is intended in this particular case. That is, rather than determining what we can infer from what is said, we try to determine what the speaker had in mind that justifies what he or she said. In terms of our notation, suppose G is a general proposition and S a specific one and

$$S \supset G$$

(that is, S implies G) is an axiom expressing a fact that a speaker and a listener mutually know. The speaker utters G in the expectation that the listener will interpret it as S. The listener must locate and use the axiom to determine the specific interpretation.

In this manner, axiom (7) provides one possible interpretation of (5), in that it specifies one of the many ways in which one thing can be *at* another, which the speaker may have meant. When a metaphorical use of *go* or *switch* or any of the other spatial predicates is encountered, axiom (7) combines with the axioms defining the spatial predicate in terms of *at* to give us the correct interpretation.

An alternative to this approach might seem to be to infer intended meaning from what was said. We would use axioms not of the form $S \supset G$ but of the form

$$G \wedge C_1 \wedge \ldots \wedge C_n \supset M$$

(that is, G together with C_1 through C_n implies M) where G is the general proposition that is explicitly conveyed, the C_i's are conditions determinable from context, and M is the intended meaning. For interpreting (5), this would require an axiom like

$$(8) \qquad at'(w, x, y) \land variable(x) \supset value'(w, y, x)$$

That is, if w is the condition of x's being at y and x is a variable, then w is also the condition of y being the value of x. To interpret (5) we would search through all axioms for axioms that, like (8), have at in the antecedent, check whether the other conjuncts in the antecedent were true, and if so, conclude that the axiom's consequent was the intended meaning. This would be equivalent to a "discrimination-net" approach to word-sense disambiguation (e.g., Rieger 1978), in which one travels down a tree-like structure, branching one way or the other according to whether some condition holds, until arriving at a unique specific interpretation at the bottom. The difficulty with this approach is that it supposes we could anticipate at the outset all the ways the meaning of a word could be influenced by context. For metaphors we would have to be able to decide beforehand on all the precise conditions leading to each interpretation. It is highly implausible that we could do this for familiar metaphors, and for novel metaphors the whole approach collapses.

As always, there are a number of inferences involving at that we would not want to draw in the case of (5). For example, in the blocks world, if BLOCK1 is at location (2,3,0), then it is impossible for BLOCK2 to be at (2,3,0) at the same time. Yet there is no difficulty whatever in two variables being "at" the same value.[6] Similarly, if a block is at a location, it is probably being held there by friction and gravity. But with variables there is no need to concern ourselves with what holds them at their values. It is probably the case in general that facts of a "topological" character lend themselves to spatial metaphors, and facts of a "physical" character do not.

---

[6]Even in our casual talk about physical reality, the inference is highly dependent on specific circumstances. We are quite comfortable saying that John and Bill are both *at* the post office.

4.3.3   A Novel Metaphor

The final example illustrates how we can represent a metaphor that depends on an elaborate analogy between two complex processes. The metaphor comes from a *Newsweek* article (July 7, 1975) about Gerald Ford's vetoes of bills Congress has passed, and is this chapter's closest approach to a literary example. A Democratic congressman complains:

(9)   We insist on serving up these veto pitches that come over the plate the size of a pumpkin.

It is clear from the rest of the article in which this appears that this means that Congress has been passing bills that the President can easily veto without political damage. There are a number of problems raised by this example, but the only ones we will address are the questions of how to represent and interpret "veto pitches that come over the plate."

The analogy here is between Congress sending a bill to the President to sign or veto and a pitcher throwing a baseball past a batter to miss or hit. Let us encode each of the processes first and establish the links between them, and then show how a natural language processing system might discover them.

A remark about notation is necessary first, however. It will be convenient to represent a sentence like "Congress sends the bill to the President" not in the most obvious way as $send(C, B, P)$, but as a statement about the existence of a condition or action SD, which is the sending by Congress of the bill to the President (cf. Davidson 1967, Hobbs 1985). We will represent this by

$$send'(SD, C, B, P)$$

The single quote may be thought of as a nominalization operator turning the sentence "Congress sends the bill to the President" into the corresponding noun phrase "the sending by Congress of the bill to the President." There are two reasons for using this notational convention: it allows us to express certain higher predications in the schemas, and it allows us to express the mapping between the schemas with greater precision. (The notation is also used in the example of Section 4.3.2, but there I thought I could slip it past the reader.)

The facts about a bill are as follows: The participants are Congress, the bill, and the President. Congress sends a bill to the President, who then either signs it or vetoes it. We will assume there is an entity C, Congress. To encode the fact that C *is* Congress, again we could write simply

$Congress(C)$.

But here also it will prove more useful to assume there is a condition, call it CC, which is the condition of C's being Congress. We will represent this

$Congress'(CC, C)$.

CC is thus the entity referred to by the noun phrase "being Congress." Similarly, there are entities B, CB, P, and CP, with the properties

$bill'(CB, B)$,

i.e., CB is the condition of B's being a bill, and

$President'(CP, P)$,

i.e., CP is the condition of P's being the President. There are three relevant actions, call them SD, SG, and VT, with the following properties:

$send'(SD, C, B, P)$,

i.e., SD is the action by Congress C of sending the bill B to the President P;

$sign'(SG, P, B)$,

i.e., SG is the action by the President P of signing the bill B; and

$veto'(VT, P, B)$,

i.e., VT is the action by the President P of vetoing the bill B. There is the condition—call it OSV—in which either the signing SG takes place or the vetoing VT takes place:

$or'(OSV, SG, VT)$.

Finally, there is the situation or condition, TH, of the sending SD's happening followed by the alternative actions OSV:

$then'(TH, SD, OSV)$.

The corresponding facts about baseball are as follows:[7] There are a pitcher x, a ball y, and a batter z, and there are the conditions cx, cy, and cz, of x, y, and z being what they are:

$pitcher'(cx, x)$
$ball'(cy, y)$
$batter'(cz, z)$

The actions are the pitching p by the pitcher x of the ball y to the batter z,

$pitch'(p, x, y, z)$;

the missing m of the ball y by the batter z,

$miss'(m, z, y)$;

and the hitting h of y by z,

$hit'(h, y, z)$.

Let omh represent the condition of one or the other of m and h occurring,

$or'(omh, m, h)$,

and th the situation of the pitching p followed by either m or h,

$then'(th, p, omh)$.

The linkage established by the metaphor is, among other things, between the bill and the ball. But it is not enough to say that B, in addition to being the bill, is also in some sense a ball, just as B has other properties, say, being concerned with federal housing loans, being printed on paper, and containing seventeen subsections. The metaphor is stronger. What the metaphor tells us is that the *condition* of B being the bill is indeed the *condition* of B being a ball. Similar links are established among the other participants, actions, and situations. That is, the baseball schema is instantiated with the entities of the Congressional bill schema, leading to the following set of propositions:

---

[7] Where individual constants, C, CC, B, ..., were used in the Congressional bill schema, universally quantified variables, x, cx, y, ..., are used here. This is because the baseball schema is general knowledge that will be applied to the specific situation involving Congress and the President. It is a collection of axioms that get instantiated in the course of interpreting the metaphor.

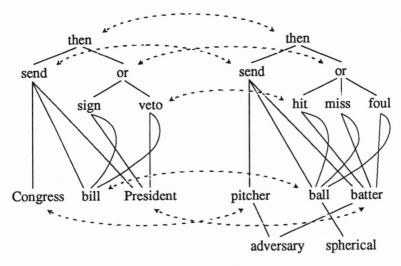

Figure 4.1 Mapping from Baseball Schema to Congress Schema.

(10) $Congress(CC, C)$ $pitcher(CC, C)$
      $bill(CB, B)$ $ball(CB, B)$
      $President(CP, P)$ $batter(CP, P)$
      $send(SD, C, B, P)$ $pitch(SD, C, B, P)$
      $sign(SG, P, B)$ $miss(SG, P, B)$
      $veto(VT, P, B)$ $hit(VT, P, B)$
            $or(OSV, SG, VT)$
           $then(TH, SD, OSV)$

The two schemas and their links are shown more graphically in Figure 4.1.

Although all of this has been described in terms of schemas, a schema in this framework is simply a collection of possibly very complex axioms that are interrelated by the co-occurrence of some of the same predicates, perhaps together with some meta-knowledge for controlling the use of the axioms in inferencing. The linkage between the two schemas does not require some special "schema-mapping" operation, but only the assumption of identity between the corresponding conditions, just as in the second example we identified "is the value of" with "is at." The difference between a conventional metaphor and a novel meta-phor is that in the case of the former the identity is encoded in an axiom like (7), whereas in the latter the identity must be drawn as an implicature. Thus, to represent the metaphor, we do not have to extend our formalism beyond what was required for the first two examples, nor indeed beyond what is required for nonmetaphorical discourse.

However, a shortcoming of this representation, as it stands, is that there is no explicit separation of the two parts of the metaphor. Thus, C is both Congress and a pitcher and P is both the President and a batter. But there is no explicit indication that the properties "Congress" and "President" belong to one side of the metaphor and "pitcher" and "batter" to the other. We could remedy this by being more careful about the difference between a condition and a description of the condition. For then we could say that the condition CC of C being Congress is identical to the condition of C being a pitcher, while the descriptions involving "Congress" and "pitcher" are distinct. We would then make assertions about the descriptions that they belong to one domain or the other. But the details of this hastily sketched idea cannot be worked out here.

No natural language processing system existing today could derive (10) from (9). Nevertheless, we can make a reasonable guess as to the basic outline of a solution. The congressman said, "We insist on serving up these veto pitches...." For someone to serve up a pitch is for him to pitch. This leads to the identification of Congress with the pitcher. To interpret the compound nominal "veto pitch," we must find the most salient, plausible relation between a veto and a pitch. From our knowledge about vetoes, we know that Congress must first *send* the bill to the President. From our knowledge about pitching, we know that for the Congress/pitcher to pitch, it must *send* a "ball" to a "batter." We have a match on the predicate "send" and on the agents of the sendings, Congress. We can complete this match by assuming, or drawing as an implicature, that the bill is the ball and the President is the batter.[8]

---

[8]Such assumptions are common in interpreting discourse. In fact, they constitute one of the principal mechanisms for resolving pronouns and implicit arguments (see Hobbs 1979).

We have almost a complete match between the two situations. The analogy will be completed when we determine which of the various possible actions that a batter can perform corresponds to the President's veto. But this is just what we need to complete the relation between "veto" and "pitch" in the compound nominal. By some means well beyond the scope of this chapter to discuss, "pitches that come over the plate the size of a pumpkin" must be interpreted to mean that the ball is easy for the batter to hit. If we assume maximum redundancy—that a veto pitch and a pitch that comes over the plate the size of a pumpkin are roughly the same thing—then we assume that the pitch is a bill/ball that the Congress/pitcher sends to the President/batter which he then finds easy to veto/hit. The analogy is complete.

As with all metaphorical expressions, as indeed with any expression, there will be a number of inferences that should not be drawn in this case—for example, that B is spherical and has stitching. But this metaphor invokes other inferences that we do accept, inferences that would not necessarily follow from the facts about the American government. It suggests, for example, that Congress and the President are adversaries in the same way that a pitcher and a batter are, and that from the President's perspective it is good for him to veto a bill Congress has passed and bad for him to sign it. What we know about the adversary relationship in baseball is vivid and unambiguous, and herein lies the power of the metaphor.

This example involves the identification of two highly structured portions of our knowledge base. It raises a question of whether our approach can handle metaphors in which one domain has much less structure, especially metaphors which impart structure to a domain that it would not otherwise have. Lakoff and Johnson (1980) demonstrate this effect by inventing a "love as a collaborative work of art" metaphor and showing some of the things that can be concluded about love as a result. I see no fresh difficulties that this would cause for my approach. Corresponding to the numerous basic links between the existing Congressional bill and baseball schemas, there would be only a few links between our knowledge of love and of collaborative works of art. If this new metaphor is productive, then corresponding to the suggestion from baseball of an adversary relationship in government, there will be numerous suggestions from the nature of collaborative works of art about the nature of love. Therefore, the effect of the new metaphor may be quite different from the effect of the ones we have examined, but the mechanisms involved in interpreting it are the same.

4.4   Some Classical Issues

4.4.1   Metaphor and Analogy

In all three examples, we have seen the same broad processes at work. They can be summarized as follows: There are two domains, which we may call the *new domain*, or the domain which we are seeking to understand or explicate, and the *old domain*, or the domain in terms of which we are trying to understand the new domain and which provides the metaphor. These are Richards' (1936) tenor and vehicle, respectively. In our examples the new domains are John's nature, computer science, and the workings of the American government. The old domains are an elephant's nature, spatial relationships, and baseball. For each old domain, we can distinguish between what may be called the *basic concepts and relationships* and *complex concepts and relationships*. For spatial relationships, "at" is a basic concept; "go," "approach," and "switch" are complex concepts. For baseball, "pitcher" and "batter" are basic, their adversary relationship is complex. In the elephant metaphor, "elephant" is basic, "has-good-memory," "clumsy" and "large" are complex. What is basic and what is complex in a particular domain are not necessarily fixed beforehand, but may be determined in part by the metaphor itself.

Each of the examples can be viewed as setting up a link between the basic concepts of a new domain and an old domain, in order that complex concepts or relationships will carry over from the old to the new. Figure 4.2 illustrates this.

To the mathematician, this diagram is familiar from Galois theory, algebraic topology, and category theory (e.g., Artin 1959, Spanier 1966, MacLane 1971). One can prove theorems in one

Old Domain		New Domain
Complex concepts and relationships	$\xrightarrow{\ 3\ }$	Complex concepts and relationships
$\uparrow 2$		$\uparrow 4$
Basic concepts and relationships	$\xleftarrow{\ 1\ }$	Basic concepts and relationships

Figure 4.2 Analogical Processes Underlying Metaphor.

domain (represented by arrow 4 in the diagram)—for example, the category of fields—by constructing a "functor" (arrow 1) to map its objects and relations into the objects and relations of another domain—for example, the category of groups—proving the theorem (arrow 2) in the second domain, and using the inverse functor (arrow 3) to map it back into the original domain.

The diagram illustrates a general paradigm for analogical reasoning. To reason in a new domain about which we may know little, we map it into an old domain, do the reasoning in the old domain, and map the results back into the new domain.

To make use of this paradigm, in our framework, for understanding the processes of metaphor, we have had to specify the nature of the links in the diagram. The horizontal links are realized by means of explicit statements like (3), or by axioms like (7) in the case of frozen metaphors, or by means of implicatures like (10) in the case of novel metaphors. The vertical links in the diagram are realized by the collections of axioms encoding the relationships between basic and complex concepts.[9,10]

But there is a problem. In category theory, once the functor maps the new domain into the old domain, then everything

---

[9]It is of course also important to specify what we mean by "domain." This issue is addressed below.

[10]Indurkhya (1986, 1987) presents an excellent formalization of metaphor and analogy as domain mapping, in which domains are viewed as theories in the logical sense and a metaphor or analogy rests on a partial function between the logical theories, from the old domain to the new domain. Many of the properties of metaphor discussed below fall out of his formal treatment. He does not embed his treatment in a larger theory of language comprehension.

we can conclude in the old domain must carry over to the new. However, in most kinds of analogical reasoning and in interpreting metaphors, only a subset of what can be concluded in the old domain will carry over to the new. The major problem for us, then, is how to determine precisely what from the old domain *does* carry over to the new. Let us elaborate on this.

There are three kinds of inferences in the old domain that must be distinguished in interpreting the metaphor.

1. The *grounds* of the metaphor, or the inferences that must be drawn if one is to make sense of the metaphor. These are what warrant the metaphor. In our first example, the grounds may be the inference that John is clumsy; in the third example, that the bill/ball is sent to the President/batter.

Black (1962) suggests a classification of theories of metaphor that includes "substitution theories," in which a metaphor is analyzed by replacing the explicit predication with those literal propositions it is intended to convey.[11] In our terms, it is the ground inferences that such theorists want to substitute for the metaphor.

2. *Disparities*, or the inferences that should not be drawn, whether because they are contradictory or irrelevant. In our examples, a disparity between John and an elephant that an elephant has a trunk, between the bill and a ball that a ball is spherical.

Richards points out that the disparities frequently play an important role: a significant effect of a metaphor may be the recognition that some of the criterial inferences that could be drawn from the explicit predication are not appropriate. The fact that John, though an elephant, is not a large animal, but a person, carries the implication that he should resemble a large animal even less. Ong (1955) suggests that a metaphor is effective only as long as it calls these disparities to mind. "John is an elephant" strikes us in a way that "the foot of a mountain" does not.

In our approach, certain disparities are considered and ac-

---

[11]Beardsley refers to this as the "literalist" theory (1958) and the "comparison" theory (1967).

tively denied, rejected when inconsistency is discovered. This active process may be compared with a cartoon in which John gradually acquires bulk, a trunk, four stocky legs while crashing along clumsily, then returns to his normal appearance. This has the flavor of a "reverse substitution" theory of metaphor, in which the *inappropriate* properties inferrable from the explicit predication, for a moment, replace the metaphor.

3. *Suggestions*, a weak term for one of metaphor's greatest powers, its suggestiveness. These are the inferences that may or may not be drawn. They are not required to interpret the metaphor, nor are they obviously inappropriate. In our first example, a suggestion is that John is large; in the third it is suggested that the President and Congress are adversaries.

There are positive and negative aspects to this suggestiveness. On the positive side, it is this more than anything else that makes metaphor such a powerful conceptual tool. We are able to draw conclusions that we could not have anticipated.

On the other hand, Lakoff and Johnson (1980) point out the dangers of mistaking the metaphor for a true description, and thus drawing too many suggested inferences without adequately examining their appropriateness. One is blinded to the limits of the metaphor, and also to alternative metaphors. Reddy (1979) discusses a specific case, the language-as-conduit metaphor and its influence on the study of communication; the theme is also developed at length by Turbayne (1962) with respect to metaphors of science.

The problem of interpreting a metaphor is to determine for the various possible inferences, which of the three classes they fall into. It is the principal thesis of this chapter that much of the solution to this problem will come from the knowledge-based interpretation processes that are already required for non-metaphorical discourse.

This position is in contrast with a commonly proposed account of metaphor interpretation. In this account, one first tries the literal interpretation, and then if that fails semantic constraints, one interprets the expression as a metaphor. That is, a separate initial step is postulated in which something is found to be wrong. There are several problems with this account. First of all, literal interpretation may not fail. Consider the following two statements

> People are not cattle.
> Whales are not fish.

Both statements are literally true biological facts. But suppose we encounter the first sentence in a political speech arguing that people cannot be herded around without consideration for their individual needs. Then it is to be interpreted as a metaphor, or if it is not a metaphor, at least it is the negation of a metaphor, and all the same interpretation processes must be called into play. Morgan (1979) gives further examples of metaphors that are or could be literally true.

A second difficulty is that all failures of literal interpretation are not due to metaphor. More often they result from metonymy, or indirect reference. For example, in

> This restaurant accepts American Express,

we are not using "accept" metaphorically as a special kind of relation between small businesses and large corporations. Rather we are using "American Express" metonymically to refer to credit cards issued by American Express. An interesting intermediate case is

> America believes in democracy.

Are we viewing America metaphorically as something which can believe, or are we using it metonymically to refer to the typical inhabitant, or the majority of inhabitants, of America?

But the principal difficulty is that this position underestimates the task of arriving at a literal interpretation of an expression. A striking example is a clause that appeared in a paper by Wallace Chafe (1980):

> Back when we were fish, ...

The intent is that this be interpreted literally, where "we" is taken to refer to all people and their ancestors indefinitely far back. But to arrive at this interpretation we have to access what we know about evolution.

An excellent example of the difficulties in interpreting literal expressions is provided by what Black (1962) calls the "compar-

ison" view of metaphor. A metaphor is seen as an elliptical form
of a simile. Thus, the metaphorical "John is an elephant" trans-
lates into the literal "John is like an elephant" or "John is like the
stereotypical elephant in certain respects." But the word "like" is
a very good example of a literal expression whose interpretation
is quite problematic. Part of the literal meaning of "A is like B"
is that A shares certain properties with B. Thus, in understanding "His house is like my house," we need to determine in which
respects the two are alike. Similarly, in interpreting "John is like
an elephant," we must discover in just what respects John is like
an elephant. But this means that the problem of interpreting
the literal "like" is isomorphic to the problem of interpreting the
original metaphor.[12]

There is generally a large overlap in the processes of literal
interpretation and metaphor interpretation, as this chapter has
argued and illustrated. Other writers have made or failed to
make this point. Searle (1979) discusses at length the difficulties
of interpreting literal utterances, but nevertheless separates these
processes from the process of interpreting the utterance once the
deviance is found, overlooking their likely identity. Rumelhart
(1979), by contrast, shows that literal interpretation is sometimes problematic, as a way of arguing for the identity of these
processes. Nunberg (1978) also argues for the identity.

Perhaps the most detailed argument is that of Miller (1979).
He shows how the interpretation of a sentence with the verb
"to be" is problematic. Even if such a sentence is used literally,
we have to determine at least whether it conveys entailment,
as in "Trees are plants," or attribution, as in "This tree is a
landmark." This can be characterized by saying that in Miller's
formula (2),

$$(2) \quad G(x) \supset (\exists F)(\exists y)(SIM(F(x), G(y)))$$

in place of SIM, there would be the relation $ENTAIL$ or AT-
$TRIBUTE$. Thus the general problem, Miller argues, is to determine which of these relations R is appropriate. That is, he

---

[12]Except of course identity is not assumed between the tenor and the vehicle. This is the standard observation about the difference between metaphor
and simile.

proposes an interpretation process in which the first step is to determine R, and then, depending on what R is, the relevant inferences are drawn.

There are two difficulties with Miller's approach. First, he does not specify how R would be determined, at least at a level of detail that would satisfy a computational linguist. It is likely that whatever processes determine that similarity is intended simultaneously determine what the similarity is. In the approach I have been presenting, the mechanisms of selective inferencing first determine what inferences should be drawn, and then it may or may not be determined what relation R best characterizes this set of inferences.

The second difficulty with Miller's approach is that it seems to imply that there is always an explicit recognition that a metaphor is being used—whenever $R = SIM$. Most examples of metaphors are not explicitly recognized as such. The reader can test this for himself: the previous paragraph depends on at least four metaphors. We have seen in this chapter that frequently the discourse operations result in a metaphor being interpreted, and that the operations themselves do not depend on the metaphor–nonmetaphor distinction. They are just the ordinary processes of deciding which inferences to draw and which to refrain from drawing.

This is not to say however that metaphors are never recognized. In many cases their recognition is just part of our general awareness of discourse, like the recognition that the speaker has used a French word, an uncommon syntactic construction, a particularly apt expression, or whatever. In other cases, the recognition might contribute to the interpretation of the sentence. For example, if someone tells me

John is a clock,

I may have to recognize explicitly that a metaphor is being used before I can get any interpretation at all. From a more computational point of view, it may be that once the grounds of the metaphor are discovered, knowledge that it *is* a metaphor often plays a role in directing further inferencing. But metaphor recognition is by no means a computationally necessary part of metaphor interpretation. It is an inference about the speaker, not the spoken.

However, not all metaphors are interpreted alike. There are various processes that might be invoked, and there are several degrees of awareness that a metaphor is being used. We can clarify this issue by using the picture of metaphor and analogy presented in Figure 4.2 to tell the life story of a metaphor. This will also throw light on another classical issue concerning metaphor: what should we count as a metaphor—is metaphor ornament or omnipresent?

The life story of a metaphor has four stages.

Think of a novel metaphor as a complex term from the old domain used in a context that requires a concept from the new domain. To interpret it we must decompose the complex term into basic concepts in the old domain, and either use available links between new and old basic concepts or surmise such links for the first time. This enables us to project the complex concept from the old to the new domain. For novel metaphors, we might expect this to require quite a bit of computing, and involve following a number of false leads.

The second stage is when the metaphor has become "familiar." The same path is followed in interpreting it, but now the salience of the required inferences is such that the computation is direct and fast. The path that had to be reconnoitered with some care when the metaphor was novel is now worn into a broad avenue that is difficult not to follow.

In the third stage, the metaphor becomes "tired." A direct link is established between the basic and complex levels in the new domain. That is, the expression acquires a new sense, it becomes technical terminology in the new domain. Nevertheless, at this stage, the metaphor can be reactivated (cf. Brooks 1965, Black 1979). We can be forced to compute anew the path whose computation is no longer ordinarily necessary. For instance, if someone tells me

I live at the foot of a mountain,

I do not see this as a metaphor. But if he then says,

Right next to the big toe.

the comparison is placed squarely before me.

Finally the metaphor dies. Because of changes in the language user's knowledge base or because of the way he learned the expression, he can not recover the path that makes sense of the metaphor. It exists only as an expression in the new domain. Yet at this stage we can still ask, as linguists, what processes "motivate" this expression in this domain (cf. Fillmore 1979)—why does the expression make sense—even though as psychologists we do not believe the person uses or could use the processes. Suppose for example someone learns the expression

set a variable *to* a value,

purely as technical terminology, without ever learning the underlying spatial metaphor of, say, setting a dial to a location. A text that would reactivate the metaphor if it were merely tired— "twist a little more" to mean "increase its value"—only baffles him. The metaphorical nature of the expression cannot be said to play a role in his interpretation of it. Nevertheless, its technical sense is not arbitrary. The technical use of "set to" was originally *motivated* by the metaphor. The processes used to interpret it when it was novel can be said to motivate it now.

In summary, the four stages can be described thus. In stage 1, the interpretation is computed. In stage 2, it is computed easily. In stage 3, it is computable, though no longer computed; at this stage, reactivation of the metaphor causes it to be computed again. In stage 4, it is neither computed nor computable, but there is nevertheless a "historical" motivation.

It is controversial whether the so-called "tired" and "dead" metaphors should count as metaphors at all, or whether we should reserve the term for novel examples. Extremes have been argued. Isenberg (1963) urges that the term "metaphor" be reserved for examples that are not just novel, but have artistic intent. Black (1979) wants to exclude the example "that no longer has pregnant metaphorical use." On the other hand, Richards (1936) and Whorf (1939[1956]) see metaphor everywhere—the "fundamental insight" of Section 4.1. On the far left, Lakoff and Johnson (1980) even view nominalizations of verbs as examples of an "event-as-object" metaphor.

Which stages are entitled to be called metaphor? Where should the line be drawn? The above account provides reasons enough for drawing the line anywhere. But in terms of the processes involved, there is simply no point in drawing a line, for they are the same at every stage. What differs is how and when they are used. The reason not to exclude the more decrepit metaphors from our investigation is that they require the same processes to be explicated as do livelier metaphors. But here the processes appear as the processes that *motivated* the expression, not the processes used to *interpret* it.

4.4.2 What Are Metaphors and Why Do We Use Them?

I have not argued in this chapter that there is no difference between metaphorical and nonmetaphorical usage. Rather I have argued that frequently the interpretation processes for both are identical. There is a distinct thing called metaphor. It is a special and very powerful way of exploiting a knowledge base in the production of discourse. This leads us to the question of what, precisely, is metaphor.

It might seem more appropriate to ask this at the beginning of a chapter on metaphor rather than at the end. But in fact what counts as a metaphor is determined by our theory of it. Of course there are central cases of metaphor—statements that are novel and literally false, function effectively in the discourse to make us see one thing in light of another, and involve a mapping between clearly distinct domains—and one's theory of metaphor must encompass these, or one is simply not talking about the same phenomenon as other writers on metaphor. But what else counts as a metaphor is theory-dependent. What one should do then is what I have done in this chapter—present the theory and then say what kinds of expressions must be considered metaphors as a consequence.

In the framework presented here, a metaphor is a linguistic expression which involves in its interpretation a mapping (computed, computable, or historical) from one domain to another via identity for the purpose of making available a new, otherwise unavailable set of inferences. Thus, "people are not cattle" and "set a variable to a value" would both count as metaphors to me.

There is still some indeterminacy in this definition, however: what is meant by "domain"? A rough first cut at this might be that a domain is a collection of predicates and axioms in a knowledge base such that the predicates are richly connected with each other by means of the axioms and are only sparsely connected with other predicates in the knowledge base. But let us look at a range of examples that illustrates the fuzziness of the notion of "domain." In

People are not cattle,

used as a political statement, we are appealing to a mapping from the domain of people and how one interacts with them, to the domain of domesticated animals and how one interacts with them. These are clearly different domains, and thus the sentence contains a metaphor. The sentence

Whales are not fish,

can also be used as a political statement in an argument against the whaling industry. Do whales and fish belong to sufficiently different domains for this to be considered a metaphor? What about

Chimpanzees are not monkeys,

in an argument against the use of chimpanzees as experimental animals? Suppose someone asks me if he can borrow one hundred dollars, and I reply

I'm not Donald Trump.

Do Donald Trump and I belong to sufficiently different domains for this to count as a metaphor?[13]

Consider another range of examples. Suppose my car is a real gas guzzler. I might say any one of the following.

My car is the Queen Mary.
My car is a tank.
My car is a truck.

The first is clearly a metaphor. The last is quite dubious. It is perhaps an argument in favor of my definition of metaphor that

---

[13]This example is due to Bob Moore.

certain fuzziness in what counts as a metaphor is reduced to the fuzziness in what counts as a domain.

In the framework presented here, we can also begin to understand why metaphors are used and why they are so pervasive. Any discourse is built on a shared knowledge base of possible inferences. By means of his utterances, the speaker triggers certain of these inferences in the listener's head. The richer the shared knowledge base, the more economical, or equivalently, the more suggestive, the discourse can be. Metaphor is a deceptively simple device for enlarging the knowledge base. By using an apt metaphor to map a new, uncertainly understood domain into an old, well-understood domain, such as spatial relationships, we gain access to a more extensive collection of axioms connecting the basic and complex levels, thereby securing a more certain grasp on the new domain conceptually and providing it with a richer vocabulary linguistically. A metaphor is good to the extent that it taps into a domain that allows a rich collection of inferences to be drawn that otherwise could not be, or equivalently, allows us to see something in a new light. When we learn a new domain, we must learn not just the logical structure of its objects, but also its basic metaphors, generally spatial, and their limits, for by this means we acquire a large chunk of knowledge about the new domain very quickly.

The interpretation problem posed by this very powerful device is that the inferences in the old domain must be sorted out properly. It has been the argument of this chapter that the ordinary context-dependent discourse operations will frequently insure that the right inferences are drawn and the wrong ones are not.

5

The Coherence and
Structure of Discourse

5.1 Discourse is Coherent

Let us begin with a fact: discourse has structure. Whenever we read something closely, with even a bit of sensitivity, text structure leaps off the page at us. We begin to see elaborations, explanations, parallelisms, contrasts, temporal sequencing, and so on. These relations bind contiguous segments of text into a global structure for the text as a whole.

Consider a specimen:

(1a)　I would like now to consider the so-called "innateness hypothesis,"

(1b)　to identify some elements in it that are or should be controversial, and

(1c)　to sketch some of the problems that arise as we try to resolve the controversy.

(2)　Then, we may try to see what can be said about the nature and exercise of the linguistic competence that has been acquired, along with some related matters.

Chomsky, *Reflections on Language*, p. 13.

Between sentence (1) and sentence (2) there is a temporal relation, indicated by "then," linking two topics Chomsky intends

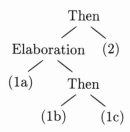

Figure 5.1 Structure of Sentences (1)–(2).

to discuss. Clause (1a) states the first topic, and clauses (1b) and (1c) elaborate on that by breaking it into two subtopics that will be discussed in sequence. This structure may be represented as in Figure 5.1. One could of course argue about details of this analysis; in fact, one of my aims in this chapter is to develop a *way* of arguing about the details.

Numerous other researchers have pointed out that such relations exist. Robert Longacre has a chapter in *Anatomy of Speech Notions* (1976) on "combinations of predications," among which he includes conjunction, contrast, comparison, alternation, temporal overlap and succession, implication, and causation. Joseph Grimes has a chapter in *Thread of Discourse* (1975) on these relations; his list includes alternation, specification, equivalence, attribution, and explanation. Others have proposed similar lists. Grimes calls these relations "rhetorical predicates," as do Mann and Thompson in their recent work (1986). Fillmore (1974) has called them "sequiturity relations." Edward Crothers (1979) calls them "logical–semantic connectives." In accord with the tradition of using idiosyncratic vterminology, I will call them "coherence relations."

The question is, what are we to make of these relations? Most authors have only pointed out their existence and listed, largely without justification, the relations most often found in texts. Longacre and Grimes describe the relations carefully. Mann and Thompson (1986) and Hovy (1988) have begun to give them more formal definitions. Crothers attempts to correlate the types of texts with the frequency of the relations that occur in them. Bonnie Meyer (1975), building on Grimes' work, classifies texts according to the structure their coherence relations impose and tries to relate that to what people remember of passages.

In this chapter a theory of coherence relations is embedded in the larger context of the knowledge-based theory of discourse interpretation sketched in Chapter 3. Section 5.2 is an account of the coherence relations, in which their intimate connection with the knowledge of the speaker and listener is explored. Of particular concern is the problem of giving formal definitions to the coherence relations in terms of inferences drawn by the listener, that would allow for the recognition of these relations.[1] In Section 5.3 it is shown how larger-scale structures in discourse are composed out of the coherence relations. This will help elucidate the elusive notions of "topic" and "genre," and allow us to examine some of the ways in which ordinary discourse is often incoherent. Thus, in Section 5.2 we examine the internal structure of the coherence relations and in Section 5.3 the structure they impose on the text as a whole. In Section 5.4 a method for analyzing discourse is suggested, which allows the structure of discourse and its underlying knowledge to illuminate each other. This method is then applied to two literary texts in Chapters 6 and 7.

5.2   The Coherence Relations

The fundamental question that must be asked about discourse is, why is any discourse longer than one sentence? That is, why do we want to call a sequence of utterances a single discourse rather than simply a sequence of utterances? What are the definitional criteria for discourse?

We may approach the problem by describing as follows the situation in which discourse between a speaker and a listener takes place. (a) The speaker wants to convey a message. (b) The message is in service of some goal. (c) The speaker must link what he says to what the listener already knows. (d) The speaker should ease the listener's difficulties in comprehension.

---

[1]Here and throughout I intend "recognition" to refer not to conscious recognition, but to the implicit or latent sort of recognition that occurs, for example, when one "recognizes" the syntactic structure of a sentence. A similar remark applies to "inference."

These considerations give rise to four classes of coherence relations. In this section I take up each of the classes in turn. For each, the coherence relations in the class are motivated by the requirements of the discourse situation. A formal definition is given for each coherence relation in terms of the inferences a listener must draw, and a number of examples, together with the relevant inferences, are given. The examples are drawn from a wide variety of sources, including an algorithm description (Hobbs 1977), a paragraph from *Newsweek* during the Watergate era (Hobbs 1976), a life-history interview with a heroin addict (Agar and Hobbs 1982), a medical textbook on hepatitis, a book in archaeology, and several other sources.

There are two places in the discussion of the examples where I may seem to be appealing to magic. I often pull facts out of the hat, saying I am pulling them out of the knowledge base; and for every plausible analysis I present, I conceal a host of other analyses that cannot be ruled out by the definitions I give. Subtheory 3 of Chapter 3, the encoding of knowledge, allows me to pull the first of these tricks, while Subtheory 6, choosing the best interpretation, allows me to pull the second. Thus, whether the tricks are indeed magic remains to be seen, but they are, at the very least, beyond the scope of this book.

5.2.1   Occasion Relation

Frequently a message is coherent because it tells about coherent events in the world. It may seem that this observation converts a hard problem into an impossible one; instead of asking what makes a sequence of sentences in a text coherent, we ask what makes a sequence of *events* in the *world* coherent. But there are a few things we can say for certain about coherence in the world. First, temporal succession is not enough. We are often puzzled by two consecutive events if we can figure out no other relation between them than mere succession, and the same is true of two sentences in a discourse:

(3)   At 5:00 a train arrived in Chicago.

At 6:00 George Bush held a press conference.

We may be able to read enough into the text to make it seem coherent, but it doesn't wear its coherence on its sleeve. When we start making assumptions to give it coherence, what criteria are we seeking to satisfy by means of the assumptions?

If we are able to see causality in the text, we are willing to conclude it is coherent. So if there is something special about the train—the maiden voyage of America's first bullet train, for example—to cause Bush to call a press conference, then the text is coherent. But causality is too strong a requirement in general. Another way of reading (6) as coherent is by assuming that George Bush was on the train and the press conference was in Chicago. In this case there is no causal relation between the two events. It is a much weaker relation, one we might call an "occasion" relation, i.e., the first event sets up the occasion for the second.

The first coherence relation is thus the *occasion* relation. There are two cases, which may be defined as in (4). In this and in all the definitions we let S_1 be the current clause or larger segment of discourse, and S_0 an immediately preceding segment. For most of the examples we may assume the "assertion" of a clause to be what is predicated by the main verb; in Section 5.3 there is some further discussion about what it is that segments assert.

(4)   **Occasion:**

1. A change of state can be inferred from the assertion of S_0, whose final state can be inferred from S_1.
2. A change of state can be inferred from the assertion of S_1, whose initial state can be inferred from S_0.

Several instances of this relation occur in the following example from a set of directions:

(5a)   Walk out the of this building.

(5b)   Turn left.

(5c)   Go to the corner.

Sentence (5a) describes a change of location whose final state holds during the event described in (5b). That location is the initial state in the change of location described in (5c). Similarly, an orientation is assumed in (5a) that is the initial state in a change of orientation described in (5b), and the final state of that

Type	(5a)	(5b)	(5c)
1	loc1 → loc2	loc2	
2		loc2	loc2 → loc3
2	angle1	angle1 → angle2	
1		angle1 → angle2	angle2

Figure 5.2 Occasion Relations in Example (5).

change is assumed in (5c). There are thus four examples of the occasion relation in this text, as illustrated in Figure 5.2. Note that there is nothing wrong with finding more than one relation between sentences. If two relations do not involve inconsistent assumptions about indeterminate material in the text, there is no harm in saying that both relations obtain.

The following are further illustrations of the *occasion* relation and rough characterizations of the inferences that need to be drawn to satisfy the definition.

(6)   Decrease N by one.
      If it is zero, reset it to MAX.

The value of the variable N is changed, and the resulting value is presupposed in the second sentence.

(7)   He noticed the broken connection in the control mechanism,
      and took it to his workshop to fix.

The first clause asserts a change in knowledge that results in the action described in the second clause.

(8)   But they commonly doubt that the message is getting through to the President,
      and now their discouragement has been compounded by the news that Nixon's two savviest political hands, Melvin Laird and Bryce Harlow, plan to quit as soon as Ford settles in.

Discouragement being compounded is a change of mental state whose initial condition is the doubt described in the first clause.

(9)   But uh you know I dropped them [goods stolen from luggage] in my pocket,

> I tied the duffel bag up and the suitcase,
> and I left it there.

Dropping the goods frees the hands for tying, and the final state of the tying holds as the speaker leaves the luggage.

Cause and *enablement* are important special cases of the *occasion* relation.

5.2.2   Evaluation Relation

The second class of coherence relations results from the need to relate what has been said to some goal of the conversation. I have called this *evaluation*. The term "metacomment" would also be appropriate. It can be defined as follows:

(10a) **Evaluation:**
> From S_1 infer that S_0 is a step in a plan for achieving some goal of the discourse.

That is S_1 tells you why S_0 was said. The relation can also be reversed:

(10b) From S_0 infer that S_1 is a step in a plan for achieving some goal of the discourse.

The discourse goal can be a very worldly goal, as in

> Did you bring your car today?   My car is at the
> garage.

From the second sentence we can infer that the normal plan for getting somewhere in a car won't work, and that therefore the first sentence is a step in an alternate plan for achieving that goal.

Frequently, the goal is a conversational goal, for example, to entertain:

> The funniest thing happened to me.
> (A story).

or

> (A story).
> It was funny at the time.

It is because of this use that I have called this relation "evaluation." An important category of conversational goals is the goal of being understood.

> ... Do you know what I mean?

Much "metatalk" is related to the rest of the discourse in this way.

This relation is close to the *cause* relation and to the *explanation* relation described below. If the state or event described in S_1 causes *the state or event described in S_0*, then S_1 *explains* S_0. If the state or event described in S_1 has caused *the speaker to say S_0*, then S_1 *evaluates* S_0.

5.2.3   Ground–Figure and Explanation Relations

The coherence relations in the third class are those directed toward relating a segment of discourse to the listener's prior knowledge. The two relations in this class are the *ground–figure* relation and the *explanation* relation.

First let us look at several examples of what one is inclined to call the ground–figure relation.

(11)   And one Sunday morning about ohhhh five o'clock in the morning I sat down in the Grand— no no, not in the Grand Central, in the Penn Station,
and while I was sitting there a young cat came up to me,...

(12)   In the round we were dancing I had barely noticed a tall, lovely, fair-haired girl they called Adrienne. All at once, in accordance with the rules of the dance, Adrienne and I found ourselves alone in the center of the circle. We were of the same height.
We were told to kiss and the dancing and the chorus whirled around us more quickly than ever.

(13)   T is a pointer to the root of a binary tree.... The following algorithm visits all the nodes of the binary tree in order, making use of an auxiliary stack A.
T1: Initialize. Set stack A empty and set the link variable P to T.

It is not sufficient to say merely that the two segments refer to the same entities, for that would not rule out pairs like (14):

(14)   Ronald Reagan was once a movie star.
He appointed George Shultz Secretary of State.

The first segment in each of the examples (11)–(13) seems to furnish background information for the second segment. It provides the "geography" against which the events of the second segment take place, or the "ground" against which the second segment places a "figure." But the "geography" can be quite metaphorical, as in example (13). Thus, a definition of the relation would be

(15) **Ground–Figure:**
Infer from S_0 a description of a system of entities and relations, and infer from S_1 that some entity is placed or moves against that system as a background.

This relation can occur in reverse order also, with the figure coming before the ground. This relation is of interest generally for causal reasons, for entities are causally influenced by the background against which they operate.

The second relation in this class is *explanation*. Its definition is as follows:

(16) **Explanation:**
Infer that the state or event asserted by S_1 causes or could cause the state or event asserted by S_0.

We don't need the inverse relation since we already have the relation *cause*. The *explanation* relation is a reason for telling a story backwards.

The following is a double example:

(17a) He was in a foul humor.

(17b) He hadn't slept well that night.

(17c) His electric blanket hadn't worked.

Sentence (17b) tells the cause of the state described in (17a), while sentence (17c) gives us the cause of (17b). In the next example,

(18) I thought well, maybe I can bum enough to get a cup of coffee and get into a movie,
'cause I was *exhausted*, I mean *exhausted*. My junk was running out.

the causality is explicitly indicated. But we would want to verify that the content is in accord with this. Exhaustion is a good reason to want shelter and, at least in the narrator's world, a

	Specific to Specific	Specific to General	General to Specific
Positive:	Parallel	Generalization	Exemplification
Negative:	Contrast	—	—

Figure 5.3 The Expansion Coherence Relations.

movie theater is shelter. Finally, consider the reported discourse

(19a)   I said, hey look you guys, why don't you just soft-pedal it.

(19b)   I said, I don't know what your story is and I care less, but you're making a general display of yourself. This place is loaded with rats. It's only a matter of time until a cop comes in and busts the whole table.

The possible undesirable consequences described in (19b) are a cause for the behavior urged in (19a).

5.2.4   Expansion Relations

The final class of coherence relations, the "expansion" relations, is the largest. These are relations that, in a sense, expand the discourse in place, rather than carrying it forward or filling in background. They all involve inferential relations between segments of the text and can probably be thought of as easing the listener's inference processes. They can be classified in terms of moves between specific and general assertions and the interaction of these moves with negation, as illustrated in Figure 5.3. I have left two blank spaces in the "Negative" row because such relations would constitute a contradiction. They might be filled in with an "*exception*" relation. One states a general truth and then gives a specific exception to it, or vice versa. But I have chosen rather arbitrarily to consider these as examples of *contrast*.

There are two important limiting cases. The *elaboration/* relation is a limiting case of the *parallel* relation; the *violated expectation* relation is a limiting case of *contrast*.

Let us consider each of these relations in turn.

The definition of the *parallel* relation is as follows:

(20) **Parallel:**

Infer $p(a_1, a_2, \ldots)$ from the assertion of S_0 and $p(b_1, b_2, \ldots)$ from the assertion of S_1, where a_i and b_i are similar, for all i.

Two entities are *similar* if they share some (reasonably specific) property. Determinations of similarity are subject to the same fuzziness and considerations of "good-ness" (Subtheory 6 again) as the coherence relations in general.

A simple example is this sentence from an algorithm description:

(21) Set stack A empty and set link variable P to T.

From each of the clauses one can infer (trivially) that a data structure is being set to a value. The predicate p is thus *set*, stack A and link variable P are similar in that they are both data structures, and the stack's emptiness and P's being equal to T are both initial conditions.

The next example is a bit more indirect. It comes from a problem in a physics textbook.

(22) The ladder weighs 100 lb with its center of gravity 20 ft from the foot,

and a 150 lb man is 10 ft from the top.

Because of the nature of the task, the reader must draw inferences from this sentence about the relevant forces. We might represent the inferences as follows:

(23) $force(100lb, L, Down, x_1)$, $distance(F, x_1, 20ft)$, $foot(F, L)$

$force(150lb, x, Down, x_2)$, $distance(T, x_2, 10ft)$, $top(T, y)$

Here the predicate p is *force*, the first arguments are similar in that they are both weights, the second and third arguments are both identical (once we identify x with L), hence similar, and the fourth arguments are similar in that they are points on the ladder at certain distances from an end of the ladder (assuming y is L). (Note that the implicit arguments x and y are resolved to L, because those implicatures, i.e., the assumption of the identities $x = L$ and $y = L$, lead to the recognition of the parallel relation and thus to the most economical and coherent interpretation.)

The next example is from a medical textbook on hepatitis:

Body Material	Contains	Concentration	Agent
blood	contains	highest concentration	HBV
semen vaginal secretions menstrual blood	contain		agent
saliva	has	lower concentrations	
(saliva of) infected individuals	in	detectable ... no more than half	HBsAg
urine	contains	low concentrations	

Figure 5.4 The *Parallel* Relation in Example (24).

(24)   Blood probably contains the highest concentration of hepatitis B virus of any tissue except liver.
Semen, vaginal secretions, and menstrual blood contain the agent and are infective.
Saliva has lower concentrations than blood, and even hepatitis B surface antigen may be detectable in no more than half of infected individuals.
Urine contains low concentrations at any given time.

The predicate p is *contain*; the diagram in Figure 5.4 indicates the corresponding similar arguments and the shared properties (the column headings) by virtue of which they are similar. Note also that the sentences are in order of decreasing concentrations; it is very frequent for particular genres or "microgenres" to be characterized by further constraints imposed on these universal coherence patterns.

The next example is from Shakespeare's 64th sonnet:

(25)   When sometime lofty towers I see down-rased
And brass eternal slave to mortal rage;

We would like to understand the chain of inferences that establish the *parallel* relation between "sometime lofty towers ... down-rased" and "brass eternal slave to mortal rage." From "down-rased" we can infer that the towers are destroyed. There are several possible interpretations of "mortal rage," but one is that mortal rage is death. To be slave to death is to be controlled by death, and thus to be destroyed. Therefore, the predicate p which each half of the parallelism asserts is *destroyed*. Next it must be determined in what way lofty towers and brass eternal are similar. Towers, being buildings, are (relatively) permanent. Brass, being metal, is relatively permanent, and if we take "eternal" to modify "brass" rather than "slave," the brass's being eternal directly implies its permanence. Thus towers and brass are similar in that they are at least seemingly permanent. These clauses are interesting also because they have an internal coherence relation of violated expectation: seemingly permanent entities are destroyed.

The next example is a Congressman's complaint about communication with the Nixon White House staff, quoted in the *Newsweek* paragraph:

(26)   We have nothing to say to Ron Ziegler,
       and Al Haig's never been in politics.

The *parallel* relation here depends on the inference from each clause that Ron Ziegler and Al Haig (similar entities, in that both were advisors to Nixon) are people with whom members of Congress cannot communicate.

Finally, an example from the heroin addict's life history:

(27)   But he *had* a really *fine* pair of *gloves*,
       and uh along with the gloves he had uh a— a cheap camera, I don't know, it was a— a *Brownie*, I think,
       and one or two other little objects that didn't amount to doodly doo.

The three clauses are in a *parallel* relation because each asserts the existence and expresses an evaluation of objects in stolen luggage.

The *elaboration* coherence relation is just the *parallel* relation when the similar entities a_i and b_i are in fact identical, for all i. It can be given the following definition:

(28)   **Elaboration:**
       Infer the same proposition P from the assertions of S_0 and S_1.

Frequently the second segment adds crucial information, but this is not specified in the definition since it is desirable to include pure repetitions under the heading of *elaboration*.

A simple illustration of the *elaboration* relation is the following:

(29a)   Go down First Street.

(29b)   Just follow First Street three blocks to A Street.

From the first sentence we can infer

(30a) $go($Agent: you, Goal: x, Path: First St., Measure: $y)$

for some x and y. From the second we can infer

(30b) $go($Agent: you, Goal: A St., Path: First St., Measure: 3 blks$)$

If we assume that x is A Street and y is 3 blocks, then the two are identical and serve as the proposition P in the definition.

A slightly more interesting case is

(31)   John can open Bill's safe.
        He knows the combination.

From the first sentence and from what we know about "can," we can infer that John knows some action that will cause the safe to be open. From the second sentence and from what we know about combinations and knowledge, we can infer that *he*, whoever *he* is, knows that dialing the combination on whatever it is the combination of will cause it to be open. By assuming that "he" refers to John and that the combination is the combination of Bill's safe, we have the same proposition P and have thus established the *elaboration* relation (and solved some coreference problems as a by-product—see Hobbs (1980)).

This example illustrates an interesting point. Some might feel the coherence relation here is really explanation. The second sentence explains the first, because knowing the combination causes one to be able to open the safe. Elaboration and explanation can blend into each other for the following reason. To recognize explanation, we need to infer S_1 *cause* S_0. To recognize elaboration, we must infer a P such that S_0 *imply* P and S_1 *imply* P. Very often the P is just the assertion of S_0 itself, that is, from each of S_0 and S_1 we infer S_0. If in addition to inferring S_0 from S_1, we recognize that we have drawn that inference, we have thereby inferred S_1 *imply* S_0. Implication can be viewed as a kind of bloodless causality; it plays the role in informational systems that causality plays in physical systems, and it seems likely to me that we understand implication by analogy with causality. That is, the inference S_1 *imply* S_0 is a variety of S_1 *cause* S_0.

It is also interesting to note that this example illustrates a very common kind of elaboration pattern that we might call *function–structure*. The first segment is one in which an eventuality is described in terms of its function in some larger environment. In the second segment the detailed, internal structure of the eventuality is described. An example of this pattern from an algorithm description is the following:

Initialize.
Set stack A empty and set link variable P to T.

The first sentence describes the role the operations play in the program as a whole. The second sentence gives the specifics of what has to be done.

The next example is from a book on the archaeology of China:

(32)   This immense tract of time is only sparsely illuminated by human relics.
Not enough material has yet been found for us to trace the technical evolution of East Asia.

From "sparse" and "illuminate" we can infer in the first sentence that the relics fail to cause one to know the "contents" of the immense tract of time. From "not enough" in the second sentence, we can infer that the material fails to cause us to know the "contents" of the technical evolution. "Relics" and "material" are the same, as are the "immense tract of time" and "the technical evolution of East Asia." The proposition P is therefore something like "The material found does not cause us to know the contents of a tract of time."

The next example is from the medical text:

(33)   Generally blood donor quality is held high by avoiding commercial donors ...
Extremely careful selection of paid donors may provide

safe blood sources in some extraordinary instances, but generally it is much safer to avoid commercially obtained blood.

Here it is crucial to recognize that blood donor quality being held high is a way of minimizing risk, which implies greater safety.

Another from the *Newsweek* paragraph:

(34)   Time is running out on Operation Candor.
Nixon must clear himself by early in the new year or lose his slipping hold on the party.

Recognition of the *elaboration* relation depends on inferring the commonality between "time is running out on ..." and "must ... by early in the new year," and then recognizing, either by knowing or assuming, that "Operation Candor" and "Nixon ... clear himself" are identical complexes of events.

Finally:

(35)   Al Haig's never been in politics—
he can't even spell the word "vote."

Both clauses are intended to imply that Haig is not knowledgeable about politics—the first by saying that he lacks the relevant experience, the second by giving an alleged example of some "political" skill he lacks.

For simplicity, in the remaining definitions, it will be assumed that the assertions of the segments that the relation links are predications with one argument. The definitions can be extended in a straightforward manner to more than one argument.

The *exemplification* relation is defined as follows:

(36)   **Exemplification:**
Infer $p(A)$ from the assertion of S_0 and $p(a)$ from the assertion of S_1, where a is a member or subset of A.

A fairly simple example is the following:

(37)   This algorithm reverses a list.
If its input is "(A B C)," its output is "(C B A)."

Recognizing the relation depends on inferring "causes X to be the reverse of X" from "reverses," inferring the causal relation between the input and output of an algorithm, recognizing that (A B C) is a list and that (C B A) is its reverse.

This more complex example is from the archaeology text:

(38)   We cannot affirm that the technical evolution of East Asia followed the same course as it did in the West.
Certainly no stage corresponding to the Mousterian tradition has been found in China.

"Cannot affirm" is matched by "no stage ... has been found." China is a part of East Asia, and "stage ... in China" is one portion of "the technical evolution of East Asia," just as the Mousterian tradition is a portion of the technical evolution of the West.

The *generalization* coherence relation is simply *exemplification* with S_0 and S_1 reversed.

There are two cases of the *contrast* relation. They can be characterized as follows:

(39)   **Contrast:**

(39a)   Infer $p(a)$ from the assertion of S_0 and $\neg p(b)$ from the assertion of S_1, where a and b are similar.

(39b)   Infer $p(a)$ from the assertion of S_0 and $p(b)$ from the assertion of S_1, where there is some property q such that $q(a)$ and $\neg q(b)$.

In the first case, contrasting predications are made about similar entities. In the second case, the same predication is made about contrasting entities.

The first example illustrates the first case:

(40)   You are not likely to hit the bull's eye,
but you are more likely to hit the bull's eye than any other equal area.

From the first clause we can infer that the probability of hitting the bull's eye is less than whatever probability counts as likely. From the second clause we can infer that the probability is greater than (and thus not less than) the typical probability of hitting any other equal area.

The second example illustrates the second case:

(41)   If INFO(M) > INFO(N), then set M to LINK(M).
If INFO(M) ≤ INFO(N), then set N to LINK(N).

What is asserted in each sentence is an implication. The first arguments of the implications are contradictory conditions. The second arguments are similar in that they are both assignment statements. Note that we must discover this relation in order not to view the instructions as temporally ordered and thereby translate them into the wrong code.

Finally, consider

(42) Research proper brings into play clockwork-like mechanisms; discovery has a magical essence.

"Research" and "discovery" are viewed as similar elements, "mechanistic" and "magical" as being contradictory. This therefore illustrates the first case.

The final coherence relation is the *violated expectation* relation, defined as follows:

(43) **Violated Expectation:**
Infer P from the assertion of S_0 and $\neg P$ from the assertion of S_1.

This is simply the first type of contrast relation in which the similar entities are in fact identical.

An example would be

(44) John is a lawyer, but he's honest.

Here one would draw the inference from the first clause that John is dishonest since he is a lawyer, but that is directly contradicted and thus overridden by the second clause.

In the following sentence from a referee's review,

(45) This paper is weak, but interesting.

one can infer from the first clause that the paper should be rejected, but from the second clause that it should be accepted.

Next:

(46) The conviction is widespread among Republicans that Mr. Nixon must clear himself by early in the new year.
But they commonly doubt that the message is getting through to the President.

Typically, if something is true of a person, that person would be expected to know it. But the second sentence denies that.

The final and most complex example is from Lenin's *State and Revolution*.

(47)   We are in favor of a democratic republic as the best form
of the state for the proletariat under capitalism;
but we have no right to forget that wage slavery is the
lot of the people even in the most democratic bourgeois
republic.

The democratic republic is best for the people under capitalism, but contrary to what one might expect from this, a rather undesirable condition—wage slavery—would still obtain.

From one perspective, we can view the coherence relations as text-building strategies, strategies the speaker uses to make the listener's comprehension easier. But that does not answer the question of why this particular set of relations should make comprehension any easier. It is tempting to speculate that these coherence relations are instantiations in discourse comprehension of more general principles of coherence that we apply in attempting to make sense out of the world we find ourselves in, principles that rest ultimately on some notion of cognitive economy. We get a simpler theory of the world if we can minimize the number of entities by identifying apparently distinct entities as different aspects of the same thing. Just as when we see two parts of a branch of a tree occluded in the middle and assume that they are parts of the same branch, so in the expansion relations we assume that two segments of text are making roughly the same kind of assertion about the same entities or classes of entities. When we hear a loud crash and the lights go out, we are apt to assume that one event has happened rather than two, by hypothesizing a causal relation. Similarly, the weak sort of causality underlying the *occasion* relation seems to be a way of binding two states or events into one. Recognizing coherence relations may thus be just one way of using certain very general principles for simplifying our view of the world.

These principles may in fact reduce to just three—causality, figure–ground, and similarity. The occasion and causal relations, explanation, and evaluation are all based on causality. The expansion relations are based on similarity. It is obvious why causality would be of interest to creatures like us, that have to maneuver our way among events beyond our control; prediction promotes survival. Our interest in figure–ground relations and similarity may reduce to causality as well. An entity (the figure) is causally influenced by the environment (the ground) in which it is located, and similar entities behave causally in a similar fashion (and when they don't, it is worthy of note). Thus, knowing these relations aids prediction.

One could argue that this style of discourse analysis is originally due to Hume. In his *Inquiry Concerning Human Understanding* (Section III), he argued that there are general principles of coherent discourse resting upon general principles for the association of ideas. "Were the loosest and freest conversation to be transcribed, there would immediately be observed something, which connected it in all its transitions. Or where this is wanting, the person, who broke the thread of discourse, might still inform you, that there had secretly revolved in his mind a succession of thought, which had gradually led him from the subject of conversation." Moreover, the three principles he proposed are very close to our own principles of causality, figure–ground, and similarity: "To me, there appear to be only three principles of connexion among ideas, namely, *Resemblance*, *Contiguity* in time or place, and *Cause* or *Effect*."

5.3   The Structure of Discourse

A clause is a segment of discourse, and when two segments of discourse are discovered to be linked by some coherence relation, the two together thereby constitute a single segment of discourse. By recognizing coherence relations between segments, we can thus build up recursively a structure for the discourse as a whole. For example, in the Chomsky passage at the beginning of this chapter, clauses (1b) and (1c) are linked by an *occasion* relation. They combine into a segment that is in turn related to clause (1a) by an *elaboration* relation. This results in a composed segment that consists of all of sentence (1); this is related to sentence (2) by an *occasion* relation. We can call the resulting structure for the text its "coherence structure." Typically, in a well-organized written text, there will be one tree spanning the entire discourse.

This notion of structure in discourse allows us to get a handle on some classical problems of discourse analysis. Here I will touch on just three: the notion of "topic," one aspect of the notion of "genre," and some of the deviations from coherence that occur in ordinary conversation.

There are really two notions of "topic" (and I refer here and throughout only to discourse topic, not sentence topic). A topic is a segment of a discourse about a single thing, and a topic is a characterization of the thing a segment is about. The first notion of topic is easy to characterize in terms of the coherence structure of texts. It is a segment spanned by a single tree which is not included in a larger segment spanned by a single tree.

There may seem to be problems with this definition when topic boundaries are uncertain. In a dialog analyzed by David Evans and me (Hobbs and Evans 1980), there is a stretch of talk about the contents of envelopes the woman is carrying, and then about her dissertation, a copy of which she is also carrying. Are there two topics—envelopes and dissertation—or just one—things she is carrying? It is hard to know what the maximal segments should be. But this uncertainty as to topic structure is exactly reflected in the uncertainty as to whether there is a *parallel* coherence relation between the two segments. Is the fact that she is carrying both the envelopes and the dissertation sufficient for the similarity required by the definition of the *parallel* relation? If so, there is one topic; if not, there are two.

The problem of characterizing the second notion of topic is a bit more difficult, and we need to back up and discuss another problem that has heretofore been glossed over. The definitions of the coherence relations are stated in terms of what utterances *assert*. In many cases it is simple to decide what is asserted: the predication expressed by the main verb. So in

The boy hit the ball.

we are asserting something like $hit(BOY_1, BALL_1)$. But there are many utterances in which this simple rule does not apply. In

They hanged an innocent man today.

it may already be mutually known that they hanged someone, and the speaker is asserting the man's innocence. Just what is asserted by single clauses may depend on the syntactic structure of the sentence, the mutual knowledge of the speaker and listener, intonation, the relation of the clause to the rest of the discourse, and other factors. This problem cannot be explored here, however. For the purposes of this discussion we will assume that the assertions of single clauses can be determined.

If the definitions of the coherence relations are to be applied to segments of discourse larger than a single clause, we need to be able to say what is asserted by those segments. We can do so if, in the composition process, when two segments S_0 and S_1 are joined by a coherence relation into a larger segment S, we have a way of assigning an assertion to S in terms of the assertions of S_0 and S_1. The assertion of S will constitute a kind of summary of the segment S.

As an approach to this problem we can divide the relations into two categories: coordinating and subordinating. Among the coordinating relations are *parallel* and *elaboration*. To recognize a coordinating relation, one must generally discover some common proposition inferrable from each segment. We can assign this common proposition as the assertion of the composed segment. For the *parallel* relation, we must infer $p(a)$ and $p(b)$, where a and b are similar by virtue of sharing some property q. We can then say that the composed segment asserts $p(x)$ where x is in $\{x \mid q(x)\}$. For example, in (22) the assertion of the whole is something like "There are downward forces acting on the ladder at some distance from an end of the ladder." In (25) the assertion is "Seemingly permanent things are destroyed." For the *elaboration* relation, we must infer some proposition P from the assertion of each segment. We can say that P is the assertion of the composed segment. In (31) the assertion is that John knows that dialing the combination will cause the safe to be open.

Among the subordinating relations are *ground–figure*, *explanation*, *exemplification* and *generalization*, *contrast*, and *violated expectation*. In these relations one of the two segments, S_0 or S_1, is subordinated to the other. We can say that the assertion of the composed segment is the assertion of the dominant segment. In fact, this is precisely what it means for one segment to be subordinated to another. In the *contrast* and *violated expectation* relations "S_0, but S_1," it is generally the second segment that is dominant, although there are exceptions. Thus, sentence (45) urges acceptance of the paper (I am happy to report). In *exemplification* and *generalization*, it is the more general statement. In *explanation* "S_0, because S_1," it is the first segment, that which is explained. For the *ground–figure* relation, the dominant segment is the figure, the segment for which the background is provided.

I'm not sure what to say about the *occasion* relation, whether to say that the composed segment asserts the assertion of the second segment, that it asserts the change, or that it asserts the occurrence of some abstract event which decomposes into the stated events, although I lean toward the last of these.

With rules such as these for assigning assertions to larger segments of discourse, it becomes easy to define the second notion of topic. A topic-in-the-first-sense is a composed segment. The topic-in-the-second-sense of this segment is the assertion assigned to it by the above rules, i.e., a kind of summary of its contents.

With this notion of discourse structure we can begin to examine conventional structures peculiar to certain genres. There are in principle many ways one could structure an account of a sequence of events, but in a given genre, for one reason or another, a few of the ways have been institutionalized or conventionalized into frozen forms. It is these constrained coherence structures that researchers who propose story grammars are seeking to characterize.

As an illustration, let us look at a conventional coherence structure for narratives that, to my knowledge, has not previously been observed. It is exhibited in the following two stories. The first is from the life-history interviews with the heroin addict:

(48a) And one Sunday morning about ohhhh five o'clock in the morning I sat down in the Grand— no no, not in the Grand Central, in the Penn Station,

(48b) and while I was sitting there a young cat came up to me, and he had his *duffel* ᵇᵘg and a suitcase, and he said, "Look," he said, "maaan," he said, "I've got to make the john. Will you keep your *eye* on the— on my stuff for me?" Well there were two . . . black fellows sitting down at the end of the line, watching this procedure, you know and I —

(48c)   for a few minutes I thought well fuck it, I — you know I'm gonna — the guy trusts me, what's the use of trying to beat him.

(48d)   But one of the black guys came over, and said, "Hey maaan, why don't you dig in and see what's there, maaan, maaan, you know, maybe we can split it,"

(48e)   and I said we're not going to split it at all, it's mine, and I picked up the suitcase, threw the duffel bag over my back and I *split*,

(48f)   and left a very irritated guy there, "I'll catch you motherfucker," he said, and I said, "well maybe you will and maybe you won't," and I'm hightailing it as fast as I can.

The second is from a life story collected by Charlotte Linde (Linde 1990).

(49a)   Uh, I started out in Renaissance studies,

(49b)   but I didn't like any of the people I was working with,

(49c)   and at first I thought I would just leave Y and go to another university,

(49d)   but a medievalist at Y University asked me to stay or at least reconsider whether I should leave or not, and um pointed out to me that I had done very well in the medieval course that I took with him and that I seemed to like it, and he was right. I did.

(49e)   And he suggested that I switch fields and stay at Y

(49f)   and that's how I got into medieval literature.

Both have the structure illustrated in Figure 5.5.

In each story, segment (a) provides background for (b). The circumstance of segment (d) causes and thus occasions the events of (e). Segments (c) and (d)–(e) are contrasting solutions. Segments (a)–(b) and (c)–(e) are related by an important subtype of the *occasion* relation—a problem and its solution. Segments

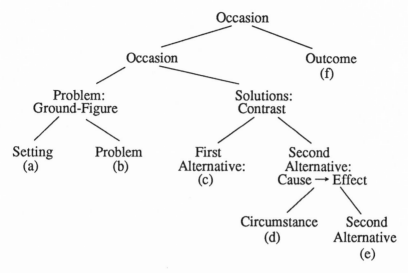

Figure 5.5 The Structure of Stories (48) and (49).

(a)–(e) and (f) are related by another important subtype—a set of events and its outcome.

It is likely that this structure is a very common pattern for stories in our culture. It is a coherence structure, but not just any coherence structure. In this convention, the *occasion* relations are constrained to be a problem–solution relation and an event–outcome relation, and the *contrast* has to be between two possible solutions.

Other genres have similar conventional constrained coherence structures. Considerations of coherence in general allow us to string together arbitrarily many parallel arguments. But it is a convention of argumentation for there to be just three, and those ordered by increasing strength. In political rhetoric, one also hears sequences of parallel statements, but for maximum effectiveness, they should be more than just the semantic parallelisms characterized by the theory of coherence. They should also exhibit a high degree of lexical and syntactic parallelism.

In a well-planned text, it is possible that one tree will span the entire text. However, conversations drift. We are likely to see a sequence of trees spanning conversational segments of various sizes, with perhaps smaller trees spanning the gaps between the

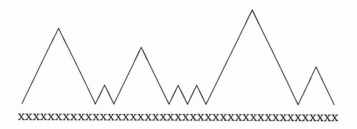

Figure 5.6 Typical Structure of a Conversation.

larger segments—something resembling what is shown in Figure 5.6. To switch metaphors in midforest, we see a number of more or less large islands of coherence linked by bridges of coherence between two points at the edges of the islands. Thus, the first sentence of a new island may be in a parallel relation to the last sentence of the previous island, but in a way that fails to develop the structure of either island. (The various ways this topic drift takes place are examined in Hobbs (1990).)

A notorious example of such local coherence and global incoherence is the phenomenon of going off on a tangent. An example of this occurs in the life-history interviews between the heroin addict Jack and the anthropologist Micheal Agar (Hobbs and Agar 1985). This interview began with Agar asking and Jack agreeing to talk about Jack's move from Chicago to New York when he was fifteen. After explaining *why* he left Chicago, Jack is now telling *how* he did it—by hitchhiking. He mentions his previous experience with hitchhiking and then slides into a reminiscence about a trip to Idaho.

(50a)   J: I had already as I told you learned a little bit about hitchhiking,

(50b)   J: I'd split out and uh two or three times, then come back,

        M: Uh huh.

(50c)   J: The one—my first trip had been to Geneva uh New York,

        M: Uh huh.

(50d)   J: And then I'd uh once or twice gone to—twice I'd gone
             to California,

(50e)   And then I'd cut down through the South,

(50f)   And I had sort of covered the United States.

(50g)   One very beautiful summer I'll tell you about some
             other time that I spent in Idaho

(50h)   to this day I remember with nothing but you know
             happiness,

(50i)   It was so beautiful,

(50j)   I'll—I'll never forget it,

(50k)   I—Right up in the mountains in these tall pine forests,

(50l)   And it was something that you know is just—it you
             know—

(50m) ⎡ J: It's indelibly in my memory,
      ⎣ M: That's   huh

(50n)   J: And nothing could ever erase it.

(50o)   M: We'll have to—we'll come back to it one day.

(50p)   J: Yeah, sometime you ask me about that.

(50q)   M: Okay.

Figure 5.7 illustrates the structure of this passage.

In utterance (50a), Jack is working out a reasonable step in
his global plan, namely, to explain that he had the means to
leave Chicago—hitchhiking. He elaborates on this in (50b) to
(50e) by giving several parallel examples of his experiences with
hitchhiking, summing up in (50f). In (50g) he gives one final
example, and here the tangent begins as he elaborates on the
beauty of the summer. In (50h) he tells of his happiness. In
(50i) he repeats that it was beautiful. In (50j) he says he'll never
forget it. In (50k) he gets specific about what was beautiful.
Utterance (50l) is probably a false start for (50m), and in (50m)
and (50n) he says again in two different ways that he'll never
forget it.

It is interesting to see how this slide happens. The crucial
utterance is (50g). Significantly, it is not clear whether it is a

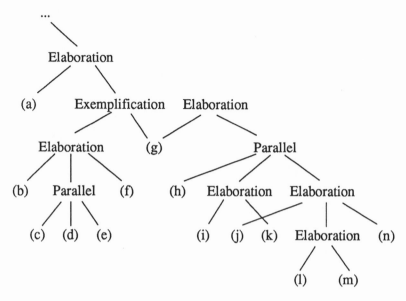

Figure 5.7 Structure of Example (50).

topicalized sentence or just a noun phrase. It is uncertain which predication is to be treated as its assertion. Insofar as it is an *exemplification* of (50f), the assertion is "I spent one summer in Idaho." But the predication that is elaborated upon subsequently, and thus functions as the assertion of (50g) from the perspective of the last half of the passage, is "The summer was very beautiful." It is the ambiguity in what (50g) asserts that enables the tangent to occur. The anthropologist finally redirects the interview in (50o) by picking up on the third predication made in (50g)—"I'll tell you about the summer some other time"—and the interview gets back on track. (This and other examples of the coherence of incoherent discourse are examined more fully in Hobbs and Agar (1985).)

This example suggests an enrichment of our view of the function of the coherence relations. The coherence relations are not merely constraints on the orderly top-down development of discourse. They are also resources to which the speaker may appeal to get him from one sentence to the next when global constraints are insufficient or insufficiently attended to. They are a means of finding a next thing to say. (See also Hobbs and Evans (1980).) A tangent occurs when there is a kind of relaxation in the discourse planning process and local coherence is pursued to the neglect of global concerns.

5.4   A Method for Analyzing Discourse

This account of the structure of discourse suggests a method for analyzing discourse. The method consists of four steps, each an order of magnitude more difficult than the one before it.

1. One identifies the one or two major breaks in the text and cuts it there. That is, one chooses the most natural way to divide the text into two or three segments. This can be done on a strictly intuitive basis by anyone who has understood the text, and among those who have understood it in the same way, there will be a large measure of agreement. This process is then repeated for each of the segments, dividing *them* in the most natural places. The process is continued until reaching the level of single clauses. This yields a tree structure for the text as a whole.

In the passage from Chomsky cited at the beginning of this chapter, for example, the major break comes between sentences (1) and (2). Within sentence (1) there is a break between the first clause and the last two, and of course a final break between the second and third clauses of the first sentence. This yields the tree of Figure 5.1.

2. One labels the nonterminal nodes of the tree with coherence relations. Proceeding from the bottom up, one devises rough accounts of what is asserted by each composed segment. Thus, in the Chomsky example, we label the node linking (1b) and (1c) with the *occasion* relation. We label the node linking the resulting segment and (1a) with the *elaboration* relation. Finally, we label the node linking (1) and (2) with the *occasion* relation.

In this step the method becomes theory-specific, as one must know what the relations are and have at least rough characterizations of them. One aid in this task is to determine what conjunctions or sentential adverbs it would be appropriate to insert. If we can insert "then" between S_0 and S_1, and the sense would be changed if we reversed the segments, then the *occasion* relation is an excellent candidate. If we can insert "because," the *explanation* relation becomes a strong possibility. "That is" or "i.e." suggests *elaboration*, "similarly" suggests *parallel*, "for example" suggests *exemplification*, and "but" suggests *contrast* or *violated expectation*. It should be emphasized, however, that these tests are informal. They do not define the relations. Conjunctions and sentential adverbials impose constraints on the propositional content of the clauses they link or modify, and in many cases these constraints are almost the same as those imposed by some coherence relation. In the best of cases there is sufficient overlap for the conjunction to tell us what the coherence relation is.

3. One makes (more or less) precise the knowledge or beliefs that support this assignment of coherence relations to the nodes. Each of the coherence relations has been defined in terms of the inferences that must be drawn from the listener's knowledge base in order to recognize the relation. When we say, for example, that an occasion relation occurs between (1b) and (1c), we have to specify the change asserted in (1b) (namely, a change in mutual knowledge about where the controversy lies, from the word "identify") that is presupposed in the event described in (1c), (the effort to resolve the controversy). Thus, we need knowledge about what change is effected by the action of identifying, and we need to know the meanings of "controversy" and "resolution" that allow us to talk about controversies being resolved.

The precision with which we specify the knowledge really can be "more or less." We might be satisfied with a careful statement in English, or we might demand formulation in terms of some logical language, embedded within a larger formal theory of the commonsense world.

4. One validates the hypotheses made in step 3 about what knowledge underlies the discourse. Agar and I (Agar and Hobbs 1982) have discussed at length how this should proceed. Briefly, one looks at the larger corpus to which the text belongs, a corpus by the same speaker or from the same culture that assumes the same audience. One attempts to construct a knowledge base or system of mutual beliefs that would support the analyses of all

of the texts in the corpus. If step 1 is a matter of minutes for a text of paragraph length, step 2 a matter of an hour or two, and step 3 a matter of days, then step 4 is a matter of months or years.

In each of these steps difficulties may arise, but these difficulties in analysis will usually reveal problematic aspects of the text. In step 1, we might find it difficult to segment the text in certain places, but this probably reflects a genuine area of incoherence in the text itself. We might find it easy to segment the text because the segments are about clearly different topics, but be unable to think of a coherence relation that links the segments. When this happens, it may be that we have found two consecutive texts rather than a single text. At times the knowledge that underlies a composed segment is not obvious, but this often leads us to very interesting nonstandard assumptions about the belief systems of the participants. For example, to justify the *explanation* relation in (18), we have to assume it is mutually understood that movie theaters are shelters. Finally, we often cannot be sure the knowledge we have assumed to be operative really is operative; looking at further data forces revisions in our assumptions.

The theory of local coherence in discourse I have sketched in this chapter is part of a larger theory that seeks to make explicit the connection between the interpretation of a text and the knowledge or belief system that underlies the text. The coherence relations that give structure to a text are part of what an interpretation is; they are defined in terms of inferences that must be drawn to recognize them, and thus specify one connection that must exist between interpretations and knowledge. The method outlined in this section can be used to exploit that connection in several ways.

Where, as in ethnography, our interest is in the belief systems, or the culture, shared by the participants, the method acts as a "forcing function." It does not tell us what the underlying beliefs are, but it forces us to hypothesize beliefs we might otherwise overlook, and it places tight constraints on what the beliefs can be.

Where our interest is primarily in the interpretation of the text, as in literary criticism, the method gives us a technique for finding the structure of the text, an important aspect of the interpretation. In placing constraints on the ideal structure of a text, it can point us toward problematic areas of the text where the ideal of coherence proposed here does not seem to be satisfied. We might ultimately decide in such cases that the ideal is in fact not satisfied, but many times we will find that the attempt to satisfy the ideal leads us to interesting reinterpretions of the whole text. The next chapter provides an example of this.

6

"Lawrence of virtuous father virtuous son": A Coherence Analysis

A sonnet is brief enough that we can examine it in detail in the framework presented in this book, with something approaching completeness. The one we shall examine is John Milton's 20th sonnet, given below. Jakobson and Jones (1970) displayed a similarly close reading of a sonnet in their brilliant analysis of Shakespeare's 129th sonnet. The difference between that analysis and this is instructive. They focused on oppositions and correspondences between various divisions of the poem as revealed by phonological, lexical, and syntactic features. At the time their analysis was written, linguistics had very little to say about how the meaning of texts was composed out of the meaning of its constituents, and in their analysis the meaning of the poem is taken for granted. Indeed, it is very nearly ignored. The oppositions that are pointed out are rarely related to meaning; they are left rather as a stunningly dense but ultimately irrelevant texture of decoration. In the last two decades substantial progress has been made in our understanding of how the meaning of texts can be represented and computed, and in the analysis below, the focus is on how this happens. No attention is given to phonological features, and lexical and syntactic features are examined only insofar as they relate to the reader's construction of the meaning of the poem. It is an account of how the ordinary meaning of the poem is accomplished, both by the writer and by the reader—an account of how the reader creates the meaning and what the writer has done to enable the reader to create it. The artistry we discover lies in the way Milton was able to exploit ordinary processes of comprehension to produce a text that conveys so much with so little.

I have broken up many of the lines of the sonnet for convenience of reference in the analysis; the fact that so many of the lines have to be broken up like this is related to a source of one of the poem's special beauties, as discussed below.

(1)	Lawrence of virtuous father virtuous son,
(2a)	Now that the fields are dank,
(2b)	and ways are mire,
(3a)	Where shall we sometimes meet,
(3b)	and by the fire
(4a)	Help waste a sullen day;
(4b)	what may be won
(5a)	From the hard season gaining;
(5b)	time will run
(6a)	On smoother,
(6b)	till Favonius reinspire
(7a)	The frozen earth;
(7b)	and clothe in fresh attire
(8a)	The lily and the rose,
(8b)	that neither sowed nor spun.
(9)	What neat repast shall feast us, light and choice,
(10a)	Of Attic taste, with wine,
(10b)	whence we may rise
(11a)	To hear the lute well touched,
(11b)	or artful voice
(12)	Warble immortal notes and Tuscan air?

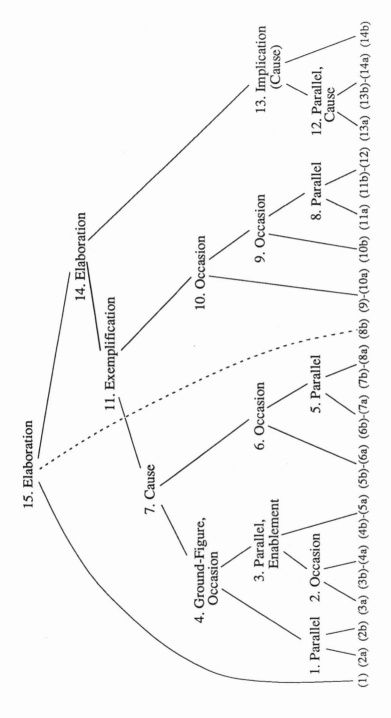

Figure 6.1 The Coherence Structure of Milton's Twentieth Sonnet.

(13a)   He who of these delights can judge,

(13b)   and spare

(14a)   To interpose them oft,

(14b)   is not unwise.

Recall from Chapter 5 that a clause forms a discourse segment, that two segments are linked by a coherence relation if inferences derivable from the content of the segments satisfy the definition of the coherence relation, and that when a coherence relation links two segments, the two together constitute a larger segment that can be summarized by what we call the assertion of the segment. In orderly discourse this yields a tree-like structure. The structure being proposed for the Milton sonnet is illustrated in Figure 6.1. In my exposition, I will attempt to justify this structure, working my way from the lower nodes to the higher ones. For convenience the nodes are numbered in the order in which they are discussed.

For the time being, we will ignore the first line. It is tempting to dismiss it altogether as a mere vocative, but as we shall see, it plays a much more important role in the poem.

The first step in a coherence analysis is to divide the text successively into intuitively perceived segments. The last thirteen lines of the poem split, conventionally enough, into two parts, lines (2)–(12) and lines (13)–(14). The first of these splits into two parts, lines (2)–(8) and lines (9)–(12). The first of these again splits into two parts, lines (2)–(5a) and lines (5b)–(8), and the first of these splits into two parts, line (2) and lines (3)–(5a).

The next step is to examine in finer detail the content of the segments to determine the coherence relations that link them and the shared knowledge or beliefs that must be called upon to establish these relations.

Lines (2a) and (2b) (Node 1) stand in a parallel relation. To establish this, we must show that the same property is being predicated of similar entities. The similar entities are the fields and the ways, similar in that both are outdoor regions. The property of being wet is inferrable both from being dank and from being mire. The definition of the parallel coherence relation is satisfied, and the summary or assertion of the composed segment, that is, of all of line (2), is "Outdoor regions are wet."[1] There is moreover an internal coherence relation in line (2b) in the contrast between "way" and "mire." The purpose of a way is to allow people to travel, and mire obstructs travel. This suggests a further parallel between lines (2a) and (2b), that these outdoor regions are difficult to work and travel in. This provides a motivation for the next segment of the discourse.

The segment from line (3a) to line (5a) splits into two parts, lines (3)–(4a) and lines (4b)–(5a). In the first of these, two clauses, line (3a) and lines (3b)–(4a) (Node 2) are linked by the occasion coherence relation. A meeting is a change of state from not being together to being together, and being together is a precondition for "[we] help waste ..." The word "help," in particular, emphasizes their being together. A summary of the composed segment (3)–(4a) might be something very close to its second clause—"Together we waste a sullen day."

Lines (4b)–(5a) stand in a parallel relation to this segment (Node 3). The two similar entities are the time periods, the day and the season, and their similarity is strengthened by their further similarly unpleasant properties: the day is sullen and the season is hard and gaining. The property asserted of the day is that together we will waste it. The property asserted (or questioned) of the season is that something will be won from it, presumably by us. Here is a place where the desire to find coherence in the text leads us to make an assumption that we would otherwise not necessarily make. If we are to recognize a parallel relation here, as is strongly suggested by the similarity of the entities, then we must see "waste together" and "win something from" as implying identical properties. We can do this by assuming that wasting time together is a good thing to do. This implicature turns out to be central to the meaning of the whole poem. One could have imagined the poet believing wasting time together was losing, not winning. A summary of the segment (3)–(5a) is then something like "Is it possible for us to waste together a sullen or hard period of time, which would be a good thing to do?"

---

[1]This loses some of the poetry of the original.

Between the highest levels of each segment there is a kind of enablement relation, in that the first questions the place for a meeting, the second questions what can be accomplished from a meeting, and the place is a prerequisite for whatever is accomplished.

Between segments (2) and (3)–(5a) (Node 4), there is at least a relation of ground–figure. Segment (2) describes the environment, and segment (3)–(5a) questions the possibility of an event that may take place within that environment. I think one could argue that there is also an occasion relation between the two segments. There is an internal contrast in segment (3)–(5a) between the unpleasant periods of time and the wasting time together by the fire, and in fact there is an implicit change of state here. We imagine the poet and Lawrence coming in from the cold rain and sitting down by the fire. The initial state of this change of state, the unpleasant weather, is precisely what is conveyed by segment (2), and we thus satisfy the definition of the occasion coherence relation.

The chain of events thus initiated is continued through line (12).

Before moving on, however, we should point out the nice ambiguity of the word "gaining." Line (2), with the "Now that" construction, implies a period of some duration. Lines (3)–(4a) make the period of a day habitual by the use of the word "sometimes," thereby indicating a period of longer than a day. Lines (4b)–(5a) mention explicitly the season, which is hard, and "gaining" adds an urgency to this, implying the season is becoming increasingly hard. But at the same time, although not supported by the syntax, "gaining" echoes the word "won," so that as the hard season is gaining in intensity, we are gaining our pleasant respite as well. Thus, the word "gaining," by itself, in its explicit function and in its echo, conveys the contrast that the first part of the poem is built around.

Lines (5b)–(6a) exhibit a skillful effect that is worth a digression. I was once shocked to see in a folded newspaper the headline

President Botha of South Africa
Refuses Nobel Peace Prize

I seized the paper and unfolded it to read on, and there was the rest of the headline.

Winner Bishop Tutu's Request for Meeting

Stanley Fish (1980, pp. 162–166) has pointed out that just such a device can be utilized in poetry. A line is broken in a place that suggests one interpretation. The reader adopts that interpretation. He or she reads on and is forced to reinterpret. The two meanings frequently create a tension that is central in the poem. Therefore, for a full appreciation of a poem or other literary work, we need to examine not just the interpretation the reader ends up with upon finishing the work. We need to look as well at the partial interpretations produced along the way, and how the author manipulates the reader through these partial interpretations. One could say, in fact, that part of what it is to read a work as an artistic production is to focus not just on the final meaning, but on the reader's experience along the road to that final meaning.

We have an example of such a device here. Immediately after the words "the hard season gaining," we see "time will run." "Time" parallels "the hard season" and "run" parallels "gaining." The reader's first impression is that the same idea is being repeated and emphasized. Not only is the hard season gaining, it is gaining quickly, and we feel a heightened sense of urgency. Then we read on and the meaning changes completely. We go from a sense of urgency to a sense of repose. But this, again, is exactly the tension that the whole first half of the poem has attempted to create.

Lines (6b)–(7a) and lines (7b)–(8a) (Node 5) stand in a parallel relation. Their elements can be lined up side by side. The "-inspire" in "reinspire" corresponds to "clothe"; both indicate a change of state from barrenness to greater fullness. The "re-" of "reinspire" corresponds to "fresh," both indicating a return to the fullness. The "earth" corresponds to "The lily and the rose" that arise from the earth. Both clauses thus describe a change of state of the earth and its products into a greater fullness.

One's first impression of clause (8b), "that neither sowed nor spun," is that it was put there for scansion and rhyme. This should make us suspicious that something else is going on. Milton is a great poet, and his words deserve to be taken seriously. However, we will postpone discussion of this clause until the end.

The word "till" indicates explicitly a relation of temporal succession between segment (5b)–(6a) and segment (6b)–(8) (Node 6). The content of the segments gives us something stronger— an occasion relation. Segment (6b)–(8) describes a change of state, whose initial state, the frozen earth, is presupposed in segment (5b)–(6a), in that it is the reason for wanting time to run on smoothly. The missing complement for the comparative "smoother" would be something like "than expected for winter." Since segment (6b)–(8) is grammatically subordinated, the composed segment (5b)–(8) asserts the same as its first constituent segment, that time will run on more smoothly.

Between segments (2)–(5a) and (5b)–(8) (Node 7) there seems to be a causal relation. Wasting time together causes time to run on smoother. Note, however, that although the causal relation is strongly implied by the word "will," it is not explicitly signalled. To recognize the relation we have to assume it is one of the poet's beliefs that spending time together eases hardship. Thus, we are again forced to draw as an implicature a proposition that is very close to the central thrust of the poem. This causal relation constitutes a good summary of the first eight lines of the poem—"Wasting time together will make time run on more smoothly than it otherwise would in a hard season."

We will analyze segment (9)–(12) by starting at the end and working backwards. Between line (11a) and lines (11b)–(12) (Node 8) there is a parallel relation. The same property, our hearing, is asserted of two entities, the lute and the voice, that are similar first in that both are musical sounds, but also in that both are characterized by artistry—"well touched" and "artful."

There is an explicit temporal relation between line (10b) and segment (11)–(12) (Node 9), and an explicit temporal relation between lines (9)–(10a) and segment (10b)–(12) (Node 10). Our desire to maximize coherence can have an interesting effect here. We would like to see these relations as not merely temporal, but as occasion relations as well. Segment (9)–(10a) describes a state of eating a meal. Segment (11)–(12) describes a state of listening to music. Segment (10b) describes a change of state from sitting at the table to standing and being ready to move to another location. It would be easy to imagine in these lines, from the words "feast," "wine," "lute," and "voice warble immortal notes," a scene of immoderate indulgence, reclining on couches feasting and drinking wine while the musicians play. The change of state described in line (10b) blocks this interpretation, by separating the eating and the music. To see the occasion relation, we need to assume that sitting in one place is appropriate for eating, and moving to another place is appropriate for listening to music.

This balance between enjoyment and control reflects in the small the balance between delight and duty urged by the poem as a whole. It is conveyed additionally by the interweaving of words of art and moderation with words of feasting. Nearly every phrase in lines (9)–(12) exhibits this internal contrastive relation. The "repast" is "neat," the "feast" is "light and choice," the "lute" is "well touched," and the "voice" is "artful." It is also significant that the cultures that are mentioned, "Attic" and "Tuscan," are cultures that are characterized by measured conduct. What is described is neither a Spartan repast nor a Roman feast.

The next coherence relation to be examined is that between segment (2)–(8) and segment (9)–(12) (Node 11). Recall that the summary of the first is "Wasting time together will make time run on more smoothly than it otherwise would in a hard season." A summary of the second is "We will enjoy a moderate but pleasant feast." Feasting together is a more specialized description of wasting time together, and enjoying and the pleasant quality of the feast are specialized descriptions of time running on smoothly. We have already at Node 7 made explicit the causal relation between wasting time together and time running on smoothly. The causal relation between the feast and the enjoyment is not explicit in segment (9)–(12), but it can be inferred—people enjoy eating—and in fact is a specialization of the implicature we drew in our analysis at Node 7. Segment (9)–(12) can thus be seen as a specialization or exemplification of segment (2)–(8).

We may also note two other progressions in the poem so far. The first is in the three questions asked. The first, in line (3a), asks about a prerequisite for meeting—the location. The second, in line (4b), asks in general terms what can be gained from the meeting. The third, in line (9), asks in more specific terms what can be gained from the meeting. This corresponds to the causal, then specialization relations we found among the segments.

The second progression is that the atmosphere continues to lighten from line (2) to line (12). We hear of ever more pleasant situations. At first, the dank and mire of the winter are described most fully, even though the sentences concern an escape from these conditions. In lines (5b)–(8) there is a balance between the harsh season and the escape from the harsh season that first spending time together and then the coming of spring will provide. In lines (9)–(12), the harsh season is gone altogether, and the pure enjoyment is tempered only by the moderation that virtue dictates. Thus in the progression of the descriptive content of the poem, we see reflected exactly the change of state that the first part of the poem is organized around.

Lines (13) and (14), unsurprisingly, provide the coda. It too has internal structure. Let us first note that the conjoined verb phrases in the relative clause of line (13a) and lines (13b)–(14a) (Node 12), stand in a parallel or perhaps even a causal relation. Two properties are being predicated of "He" and "these delights," first "judge" and second "spare to interpose oft." They are similar in that they are both propositional attitudes one takes toward actions. Thus, we have at least a parallel relation between the clauses. However, one can argue that there is a causal relation as well, since judging whether or not to take an action is a prerequisite for taking it. This rule is not necessarily among the reader's own beliefs. There are, after all, many precipitous people who do not judge before they act. But we may assume it is a belief of the poet's, and if we do, we can discover the stronger, causal coherence relation between the clauses. Once again, to maximize the coherence of our interpretation, we are driven to draw as an implicature a rule that is very close to the central meaning of the poem.

Lines (13)–(14a) constitute the subject of a sentence and line

(14b) provides the predicate complement (Node 13), so in one sense there is no discourse structure to analyze. It is all in the syntax. But there is an internal coherence relation between the relative and main clauses. Parallel mental qualities are being predicated of "He," in lines (13)–(14a) an attitude toward one's actions, and in line (14b) a general mental and moral characteristic. Moreover, when we say "He who VP_1, VP_2," we are asserting the general rule $(\forall x)\,VP_1(x) \supset VP_2(x)$. That is, for all entities x, if the verb phrase VP_1 is true of x, then so is the verb phrase VP_2. So the relation is more than simply parallelism, it is implicational (or, as I suggested in Chapter 5, causal).

The same balance between words of enjoyment and words of moderation that characterized segment (9)–(12) also characterizes segment (13)–(14). "Delights" is balanced by "judge." Rather than using the bare adjective "wise," Milton softens it to "not unwise," thereby balancing the literal predication with the ephemeral suggestion of "unwisdom." The use of the word "interpose," rather than a phrase like "indulge in," for example, by itself suggests moderation. It forces us to focus on those activities, presumably the more serious business of life, among which the delights are interposed. Finally, the word "spare" by its own ambiguity suggests this balance so central to the meaning of the poem.

The intended meaning of the word "spare" has been a matter of controversy throughout the critical history of the poem, as summarized in *A Variorum Commentary on the Poems of John Milton* (Woodhouse and Bush (1972), pp. 474–476). One group of critics argues that the phrase "spare to interpose" is to be interpreted as "refrain from interposing," whereas the other group of critics argues that it means "spare time to interpose." The difference is, of course, complete. It is a question of whether the principal thrust of the poem is one thing or its opposite. Stanley Fish ((1980), pp. 148–152) uses this controversy itself as evidence for an intended ambiguity. The analysis given here is quite similar to Fish's. But contrary to Fish, I will argue that the readings are not actually contradictory, but merely indicative of the moderation that is the pervading message of the entire sestet.

Let us heed Fish's advice and examine our experience as we read carefully and attempt to construct coherent interpretations of successive initial segments of the poem. When all that has been seen is "... of these delights can judge, and spare," only one interpretation is possible. "Judge and spare" constitutes a conjoined verb phrase sharing the object "these delights." The sense must be "refrain from." This is reinforced by the adjectival meaning of "spare" as thin and lean, a counterpoint to the word "delights," not supported by the syntax, but perhaps primed in the reader's mind. However, the comma (and in the original version of 1673 the capitalization of "And") militate against parsing the words as conjoined verbs. Whether or not this meaning occurs to the reader at this point, it could not survive long.

We continue on to the words "To interpose them." At this point, the "refrain from" meaning cannot possibly be correct. It would be in flat contradiction to the rest of the poem. It would perhaps be possible to resolve the contradiction, but the resulting interpretation—one must abstain from even moderate delights—would be quite jarring, in contrast to the easy, relaxed, and moderate tone of the entire sonnet. We are thus forced toward the other sense of "spare." We are being urged to spare time to interpose the delights, and this meaning meshes well with and in fact sums up the entire poem.

Then we come to the word "oft." Suddenly the "refrain from" sense becomes possible again, but this time in a way that does not contradict our previous interpretation. In any sentence in which an adverbially modified clause is embedded in a higher operator, there is a question of what the argument of the higher operator is. It is always possible for it to be the adverbial rather than the main verb. In "He has not written any papers recently," it is not his writing of papers that is being denied, but the *recency* of any such writing. Similarly here, it is not interposing delights among our duties that we should refrain from, but doing so too often. We can take Milton to be saying that we should spare time to interpose delights, but we should refrain from interposing them too often. Both senses of "spare" can therefore be adopted, not in a contradictory but in a qualifying fashion.

This analysis differs somewhat from Fish's. Fish argues that a sharp and irresolvable ambiguity was intended by Milton and was his way of throwing the whole moral problem of duty versus delights back to the reader. I find this reading more clever than plausible. In the reading suggested here, the ambiguity is also intended by Milton, but not as a way of setting up an irresolvable conflict. Rather, it is a way of urging a measured approach to the moral problem, one that allows for the proper amounts of each activity.

In the ambiguity of the single word "spare" we thus see an example of something that is not uncommon in the best literature. The rhetorical device of paradox asserts something that is contradictory. In the rhetorical device of irony, a contradiction is implicit; what is said conflicts with what can be taken to be a belief shared by the speaker and listener. We can take as the purest example of these devices the ordinary utterance, "Well, ... that's true and it isn't true." We do not want to convict the speaker of this utterance of inconsistency, and this forces us to reinterpret the utterance, or interpret it more deeply. The utterance is thus a way of saying, "The situation is not so simple."[2] We interpret it by inferring something like "It is true in that P_1, P_2, and P_3, but it is not true in that Q_1, Q_2, and Q_3," and this elaboration is just the kind of more complex analysis of the situation that is being urged. The devices of paradox and irony are, in their highest uses, ways of conveying a complexity that is otherwise difficult to convey with the sometimes too blunt instrument of language. Milton has used the ambiguity of the word "spare" to the same effect. It is not the case simply that one should indulge in delights or that one should not indulge in delights. The situation is more complex. One must judge carefully and enjoy the delights in moderation.

The next coherence relation we need to examine is that between segment (2)–(12) and segment (13)–(14) (Node 14). The assertion of the first is "Wasting time together will make time run on more smoothly than it otherwise would in a hard season." The assertion of the second is "If one engages in these delights

---

in moderation, one is not unwise." These stand in an elaboration relation. The same principle is being communicated by both segments. The first does so in a way that describes explicitly the beneficial consequences of wasting time together. The second does so by describing a character trait, wisdom, that is possessed by one who takes actions that have beneficial consequences.

There has been a moral tone throughout this poem, but it has been in the lexical balance we have discussed, rather than in what has been asserted. There is no necessarily moral implication in the idea that it is not unwise to waste time with friends occasionally in a hard season because it makes time run on more smoothly. In the topmost coherence relation in the poem, that between line (1) and lines (2)–(14) (Node 15), the moral character of the message becomes explicit. Line (1) becomes more than a mere vocative if we take it to be predicating virtue of Lawrence, if we take Lawrence to be one of the possible he's in line (13a), and if we take the wisdom of wasting time together in moderation to be one part of virtue. Segment (2)–(14) then becomes an elaboration (or possibly a specialization) of line (1). One aspect of the detailed nature of virtue is being explicated. The poem is, of course, an exhortation to Lawrence, first recognizing his virtue, to take a milder, but not too mild, view of virtue.

Finally, let us return to the one unaccounted for phrase, in the middle of the sonnet, at the end of the octet, "that neither sowed nor spun." This is an allusion to Matthew 6:26,

> Behold the fowls of the air: for they sow not, neither
> do they reap, nor gather into barns; yet your heavenly
> Father feedeth them. Are ye not much better than
> they?

and to Matthew 6:28,

> And why take ye thought for raiment? Consider the
> lilies of the field, how they grow; they toil not, neither
> do they spin:

in the Sermon on the Mount. Jesus taught that life is not all work directed toward sustenance, and Milton's sonnet urges the same. (There is a significant difference, however; in this passage Jesus was telling this followers to take more time not for moderate delights but for seeking after God.) By means of this allusion, Milton claims the sanction of religion for what he is advocating.

Many, perhaps all, poems work by presenting us with a very rich fabric of coreference (very abstractly defined, to include such things as alliteration and rhyme) and inviting us to discover the coherence. Very often, discovering the coherence requires us to call forth and force into combination large, highly structured conceptual schemas that are heavily charged emotionally.[3] A fairly pure example can be seen in Ezra Pound's famous haiku, "In a Station of the Metro":

> The apparition of these faces in the crowd;
> Petals on a wet, black bough.

Here two powerful but unrelated images are presented to us individually and we are forced to discover their relation. In Milton's sonnets generally and in the twentieth one in particular we are not given such large-scale, separate, and unitary images to make of what we can. The poems seem to proceed on a much more literal level. Yet this poem is also built around a highly charged tension, that between hard duty in the face of adversity and the easy comforts of relaxed fellowship. The tension operates in a much more interwoven fashion than it does in Pound's haiku, but just as in the haiku, juxtaposition seems to promise coherence and thus impels us to try to construct a coherence. In Milton's sonnet, we find that when we have constructed the coherence, we have done so by bringing together schemas for duty and delights and by recognizing causalities and resolving conflicts between them. For example, at Node 3 we were led by the similarity of entities to expect a parallel coherence relation between the constituent segments, and thus led to draw as an implicature the

---

[3]This characterization is similar to an account of the aesthetic experience proposed by Bever (1986). In his view an aesthetically satisfying experience is one that "stimulates a conflict in perceptual representations, which is resolved by accessing another representation that allows the two conflicting ones to coexist" (p. 316).

proposition that moderate delights can be good. We were required to draw essentially the same implicature to recognize the causal relation at Node 7. Resolving the clash between the words of indulgence and the words of control in the sestet, and especially grappling with the ambiguity of the word "spare," forced us into a more complex appreciation of the delicate relationship between duty and delight.

The coherence analysis of a text is a way of reading closely. It is a way of forcing one's self to ask certain questions about how the various elements of the text fit together and why. Much of the time it forces us to make explicit what we recognized implicitly on casual reading. In other cases, it leads us to discover new beauties in the work that we would otherwise have missed.

After this kind of microanalysis, however, one must always read the poem through one more time, to experience it holistically, informed, however, by the subtle beauties one has discovered. To encourage that, I close this chapter with the entire sonnet, this time without the referential apparatus.

Lawrence of virtuous father virtuous son,
> Now that the fields are dank, and ways are mire,
> Where shall we sometimes meet, and by the fire
> Help waste a sullen day; what may be won

From the hard season gaining; time will run
> On smoother, till Favonius reinspire
> The frozen earth; and clothe in fresh attire
> The lily and the rose, that neither sowed nor spun.

What neat repast shall feast us, light and choice,
> Of Attic taste, with wine, whence we may rise
> To hear the lute well touched, or artful voice

Warble immortal notes and Tuscan air?
> He who of these delights can judge, and spare
> To interpose them oft, is not unwise.

7

Structuring in Nerval's
Sylvie

WITH PATRIZIA VIOLI

7.1 Introduction

This chapter has two aims—first, to apply and extend the method
of text analysis presented in Chapter 5 to a longer literary work,
Gerard de Nerval's *Sylvie*, and second, to use this analysis as a
way of explicating the structure and the meaning of the novella
and the close relationship between the two.

Implicatures are central to our analysis. An *implicature* is a
proposition the reader or listener assumes to be a belief shared
with the speaker or writer in order to maximize the coherence
of the interpretation of the discourse. We saw an implicature
in Section 4.3.3 function in the interpreting of a complex metaphor. We saw a number of cases of implicature in the examples
of Chapter 5, leading for instance to the resolution of pronouns
and omitted arguments. The interpretation of Milton's sonnet in
Chapter 6 required several implicatures central to the meaning of
the poem. We will see the same thing happening in the analysis
of this work, for much of the coherence of *Sylvie* depends not on
knowledge the author and reader already share, but on assumptions the author wishes the reader to make about the deeper significance of events recounted or about what particular concrete

entities symbolize. Thus, in Section 7.2.1 we see how recognizing the structure of the novella depends on our assuming, or drawing the implicature, that Sylvie and Aurélie represent reality while Adrienne represents a romantic ideal. Coherence theory gives us a way of validating such interpretations of a text in terms of text structure.

One difficulty we face is that our style of microanalysis requires a great deal of time-consuming attention to detail. It is simply not feasible to perform it on an extended text, literary or otherwise. Hence one of the goals of this work is to find reasonable principles of selection of passages for microanalysis, so that what we learn about these will tell us about the text as a whole. We have chosen two sorts of passages. In Section 7.2 we look at the level of the story itself, examining four key episodes. In Section 7.3 we consider the level of the narration, examining the sometimes confusing transitions in the story from one time to another. In their different ways, both these sets of passages tell us something about a core theme of *Sylvie*—the narrator's failure to integrate romance and reality, the past and the present.

In this chapter, as in Chapter 6, the intent is to examine not merely the meaning of the text, but much more, how the writer and reader each accomplish the meaning of the text.

7.2   The Failure of the Romantic Image

7.2.1   The Global Structure of Sylvie

Sylvie is a story of the narrator's romantic relationships with three women—a girl Adrienne he saw singing one evening at a festival in his childhood, a peasant girl Sylvie whom he grew up with, and an actress Aurélie whom he becomes involved with as an adult. The story is told in fourteen chapters that introduce, maintain and finally "resolve" the story's fundamental tension between the narrator's romantic image of Adrienne and ordinary reality. The events of the story take place at six times which we call T0, T1a, T1b, T2, T3 and T4.

The story begins at time T2 with the narrator N as an adult. (We will refer to the narrator as N, to the author as Nerval.) N leaves the theater, infatuated by the actress Aurélie. After dining with friends, he happens to read in a newspaper of an annual festival at Loisy, a village of his childhood.

In Chapter 2 he has gone to bed, and in a half-dream, we are brought back to time T0, in N's childhood. At a festival at an unnamed chateau, he sees Adrienne, the daughter of a local noble, for the first time. She is called upon to perform a dance and does so with angelic, ethereal grace. The memory of this event influences N's relations with the other women he knows for the rest of his life.

In Chapter 3 he suddenly decides to go to the festival at Loisy, gets up again at one o'clock in the morning, and catches a coach.

In Chapters 4 through 6 we find N at a festival in the country, the festival that in Chapter 3 he intends to return to. It is common for the reader to confuse the time of these events on first reading. They seem to be a continuation of the events of Chapter 3. But as we learn later (or immediately, if we interpret correctly), they are really the memory of an earlier return, at a time we call T1a. On that visit, N meets Sylvie at the festival in the evening. The next morning they take a country walk together, ending up at the cottage of Sylvie's aunt in the village of Othys, where they dress themselves in the aunt's and uncle's wedding clothes. We learn that Sylvie has also been the subject of N's romantic fantasies.

Chapter 7 first returns us to time T2, with N on the road at night going toward Loisy. He slips into his only other memory of Adrienne, this one of an even more dreamlike character. N and Sylvie's brother are on an evening excursion at a time we will call T1b (since it is unclear whether it precedes or follows time T1a.) They happen upon a ceremony of sorts in which Adrienne is portraying an angelic spirit in a mystery play at the Abbey of Châalis. This chapter advances and elaborates on N's romantic image of Adrienne.

At this point we have been introduced to the three women. We have a sense of the power in N's life of the romantic image of Adrienne. She has defined a role in his life. We are not sure what that role is; Adrienne's dramas are undoubtedly symbolic, but symbolic of what? Yet we have seen that both Sylvie and

Aurélie are candidates for that role. Moreover, we sense that either woman will become significant to him only insofar as she fills that role. Thus the tension has been set up that the second half of the story resolves.

In Chapters 8 through 12, N has returned to Loisy at time T2 and the romantic involvement with Sylvie receives its final resolution, solidly on the side of ordinary reality. N sees Sylvie at the ball. The next day they walk through the countryside together, ending up at Châalis. There in a climactic scene, N has Sylvie try to sing as Adrienne had in the same place. Sylvie fails utterly to evoke the same image. N decides that Sylvie can never be more than a sister to him. The next day N learns that Sylvie is engaged to his foster brother, a coarse sort of country fellow.

In Chapter 13 the time suddenly begins to advance very quickly. During a time T3, N and the actress Aurélie become lovers. Some months later, N and Aurélie are at the unnamed chateau together, where he tells her about Adrienne. Aurélie dismisses him out of hand, realizing now that he loves her only insofar as he mistakes her for Adrienne. Shortly thereafter, she decides to leave him for another man. Reality has failed to live up to the romantic image of Adrienne, and that romantic image has destroyed N's relationships with the two real women in his life.

In the final chapter, during some later period T4, N revisits Loisy. There in a final triumph of ordinary reality over the romantic image, we see a Sylvie who is a housewife and a mother, it is suggested that the resemblance between Aurélie and Adrienne was illusory, and we learn that Adrienne herself had died some years before. N tells us he now believes the romantic image to have been a delusion, but he fails to convince us that he really believes it.

The structure of the story is illustrated in Figure 7.1. The nodes of the tree are labelled with the relations that obtain between the segments of the text they subsume.

This is perhaps an appropriate place to make an important point. Although Figure 7.1 looks like a purely syntactic description of the structure of the story, it is also deeply semantic. The relations have definitions in terms of what has to be assumed in the shared complicity between the writer and the reader. In this

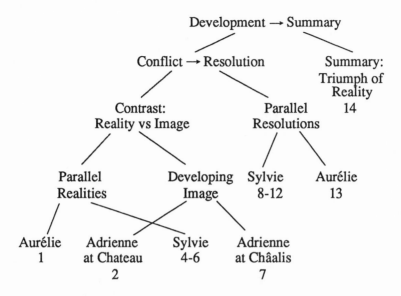

Figure 7.1 Global Structure of *Sylvie*.

case for example, to arrive at the structural interpretations of the stories of Sylvie and Aurélie as parallel exemplifications of reality in contrast with the romantic image of Adrienne, we have to make assumptions about what the three women symbolize; conversely, it is in part because of our desire to see the story as a unified, coherent whole, that we make these assumptions.

The four "singing" scenes—two with Adrienne, one with Sylvie, and one in which we would claim Aurélie fails to sing—occupy key climactic positions in this structure and exhibit revealing similarities and differences. A close examination and comparison of these episodes should therefore lead us to a deeper understanding of the nature of N's romantic image and the failure of the other women to instantiate it.

7.2.2   Adrienne at the Chateau

The first scene to be analyzed begins as follows:

1.1   A peine avais-je remarqué, dans la ronde où nous dansions, une blonde, grande et belle, qu'on appelait Adrienne.

In the round we were dancing I had barely noticed a tall, lovely, fair-haired girl they called Adrienne.

1.2	Tout d'un coup, suivant les règles de la danse, Adrienne se trouva placée seule avec moi au milieu du cercle.	All at once, in accordance with the rules of the dance, Adrienne and I found ourselves alone in the center of the circle.
1.3	Nos tailles étaient pareilles.	We were of the same height.
1.4	On nous dit de nous embrasser, et la danse et le chœur tournaient plus vivement que jamais.	We were told to kiss and the dancing and the chorus whirled around us more quickly than ever.
1.5	En lui donnant ce baiser, je ne pus m'empêcher de lui presser la main.	As I gave her this kiss I could not resist pressing her hand.
1.6	Les longs anneaux roulés de ses cheveux d'or effleuraient mes joues.	The long tight curls of her golden hair brushed my cheeks,
1.7	De ce moment, un trouble inconnu s'empara de moi.	and from that moment on an inexplicable confusion took hold of me.[1]

The episode is described in sentences (1.1) to (1.6), with (1.4) describing the key event—kissing and dancing. Sentence (1.1) introduces the key character, Adrienne, and (1.2) sets up the situation. (1.3) adds information that allows us to better visualize the key event; moreover, the detail is one that promotes N's idealization of Adrienne. Sentences (1.5) and (1.6) elaborate on the key event by giving small but, to N, significant details—details of their physical contact, perhaps symbolic for N of the spiritual contact he desires. Finally, the effect of the episode, a very long term effect, is stated in clause (1.7). The entire story is about N's subsequent attempts to deal with this inexplicable confusion. This structure is illustrated in Figure 7.2.

The next paragraph deals with Adrienne's singing. We discuss it in detail since this is the scene that is repeated in the subsequent key scenes to be analyzed.

1.8	La belle devait chanter pour avoir le droit de rentrer dans la danse.	The girl had to sing a song in order to regain her place in the dance.

---

[1] All the translations in this chapter, except (6.7)–(6.9), are from *Selected Writings of Gérard de Nerval* (1957), translated by Geoffrey Wagner.

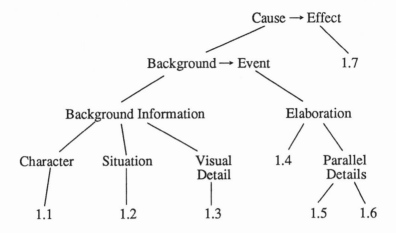

Figure 7.2 The Structure of (1.1)–(1.7).

1.9 On s'assit autour d'elle,	We sat around her
1.10 et aussitôt, d'une voix fraîche et pénétrante, légèrement voilée, comme celle des filles de ce pays brumeux,	and straight away, in a fresh, penetrating, slightly filmy voice, like a true daughter of that misty region,
1.11 elle chanta une de ces anciennes romances pleines de mélancolie et d'amour, qui racontent toujours les malheurs d'une princesse enfermée dans sa tour par la volonté d'un père qui la punit d'avoir aimé.	she sang one of those old ballads, full of melancholy and love, which always tell of the sufferings of a princess confined in a tower by her father as a punishment for having fallen in love.
1.12 La mélodie se terminait à chaque stance par ces trilles chevrotants que font valoir si bien les voix jeunes, quand elles imitent par un frisson modulé la voix tremblante des aïeules.	The melody ended at each stanza in those wavering trills which show off young voices so well, especially when in a controlled tremor, they imitate the quavering tones of old women.

Sentences (1.8) and (1.9) explain and set up the situation. Then (1.10) to (1.12) describe three aspects of the singing—the quality of Adrienne's voice, the subject matter of her song, and her wavering trills. It will be of interest to see how these aspects are treated in the subsequent singing episodes. It is perhaps also significant that the subject matter of her song could be what N, as a boy at T0, imagines about, and certainly for the reader prefigures, Adrienne's subsequent confinement in a convent.

The next paragraph does the most to create the dreamlike mood associated with the event.

1.13 A mesure qu'elle chantait,	As she sang,
1.14 l'ombre descendait des grands arbres, et le clair de lune naissant tombait sur elle seule,	the shadows came down from the great trees, and the first moonlight fell on her as she stood alone
1.15 isolée de notre cercle attentif.	in our attentive circle.
1.16 Elle se tut,	She stopped,
1.17 et personne n'osa rompre le silence.	and no one dared to break the silence.
1.18 La pelouse était couverte de faibles vapeurs condensées, qui déroulaient leurs blancs flocons sur les pointes des herbes.	The lawn was covered with thin veils of vapor which trailed white tufts on the tips of the grasses.
1.19 Nous pensions être en paradis.	We imagined we were in paradise.
1.20 Je me levai enfin, courant au parterre du château, où se trouvaient des lauriers, plantés dans de grands vases de faïence peints en camaïeu.	Finally I got up and ran to the gardens of the chateau, where some laurels grew, planted in large faience vases with monochrome bas-reliefs.
1.21 Je rapportai deux branches, qui furent tressées en couronne et nouées d'un ruban.	I brought back two branches which were then woven into a crown and tied with a ribbon.
1.22 Je posai sur la tête d'Adrienne cet ornement,	This I put on Adrienne's head
1.23 dont les feuilles lustrées éclataient sur ses cheveux blonds aux rayons pâles de la lune.	and glistening leaves shone on her fair hair in the pale moonlight.
1.24 Elle ressemblait à la Béatrice de Dante qui sourit au poète errant sur la lisière des saintes demeures.	She was like Dante's Beatrice, smiling on the poet as he strayed on the verge of the blessed abodes.

This paragraph exhibits an interesting interwoven structure. Three themes are repeated again and again. First are the bare events—Adrienne sings in (1.13), she stops in (1.16), and in

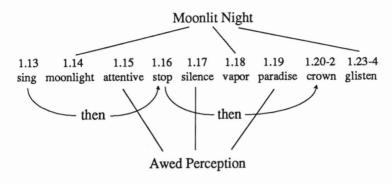

Figure 7.3 The Structure of (1.13)–(1.24).

(1.20) to (1.22) N brings her a laurel crown. But interspersed with these events, we have the awed perception of Adrienne that N attributes not just to himself but to the whole gathered assembly—(1.15), (1.17), and (1.19). These increase in intensity, as the crowd is attentive, then speechless, then imagines itself in paradise. Finally, there is interspersed the description of the moonlit night—in (1.14), (1.18), and finally in (1.23) and (1.24)—primarily as it lends its ghostly beauty to the figure of Adrienne.

Figure 7.3 illustrates this structure.

The episode ends quickly.

1.25	Adrienne se leva.	Adrienne rose.
1.26	Développant sa taille élancée, elle nous fit un salut gracieux,	Showing off her slender figure she made us a graceful bow
1.27	et rentra en courant dans le château.	and ran back to the chateau.

7.2.3 Châalis

The only other time N sees Adrienne is in a similar circumstance, equally romantic and dreamlike, but differing in significant ways. He and Sylvie's brother come upon a country ceremony one evening where a mystery play is being performed.

2.1	Ce que je vis jouer était comme un mystère des anciens temps.	What I saw performed was like a mystery play of ancient times.

2.2	Les costumes, composés de longues robes, n'étaient variés que par les couleurs de l'azur, de l'hyacinthe ou de l'aurore.	The costumes were long robes, varied only in their colors, of azure, hyacinth, and of the color of dawn.
2.3	La scène se passait entre les anges, sur les débris du monde détruit.	The action took place among angels, on the ruins of the shattered world.
2.4	Chaque voix chantait une des splendeurs de ce globe éteint, et l'ange de la mort définissait les causes de sa destruction.	Each voice sang one of the splendors of this vanished world, and the angel of death declared the causes of its destruction.
2.5	Un esprit montait de l'abîme, tenant en main l'épée flamboyante,	A spirit arose from the abyss, holding in its hand a flaming sword,
2.6	et convoquait les autres à venir admirer la gloire du Christ vainqueur des enfers.	and summoned the others to come and adore the glory of Christ, the conqueror of hell.
2.7	Cet esprit, c'était Adrienne transfigurée par son costume, comme elle l'était déjà par sa vocation.	This spirit was Adrienne, transfigured by her costume as she already was by her vocation.
2.8	Le nimbe de carton doré qui ceignait sa tête angélique nous paraissait bien naturellement un cercle de lumière;	The halo of gilt cardboard around her angelic head seemed to us, quite naturally, a circle of light;
2.9	sa voix avait gagné en force et en étendue,	her voice had gained in strength and range,
2.10	et les fioritures infinies du chant italien brodaient de leurs gazouillements d'oiseau les phrases sévères d'un récitatif pompeux.	and the endless *fioriture* of Italian singing embroidered the severe phrases of stately recitative with their bird-like trills.

Sentences (2.1) and (2.2) set the scene. Sentences (2.3) and (2.4) describe the initial action against which the figure of Adrienne will contrast. Concerning Adrienne we are then told about the same aspects as in the previous singing episode, but there are significant differences. In (2.5) she appears suddenly in the center of the circle as she had before, but now out of the abyss and not out of the circle of dancers. We are told the subject matter of her song in (2.6); this time instead of a song of imprisonment it is a song of salvation. We are told about the quality of her voice in (2.9), no longer slightly filmy but stronger. We are again told of the trills (2.10), but rather than the natural trills of youth, they exemplify a studied style. Just as before the assembly viewed her with awe as the moonlight bathed her, now ordinary cardboard seemed a circle of light. She is no longer someone whom N may kiss, but is now transfigured into something infinitely distant. Whereas in the previous singing scene Adrienne was someone almost within reach, here she is utterly beyond N's grasp. She is not merely an object of romantic love; she has become an object of religious adoration as well.

The next paragraph is interesting because it questions the epistemic status of the entire episode. It maintains a fine tension between dream and reality, reflecting in the small a primary theme of the entire story.

2.11 En me retraçant ces détails, j'en suis à me demander s'ils sont réels, ou bien si je les ai rêvés.

As I retrace these details I have to ask myself if they were real or if I dreamed them.

2.12 Le frère de Sylvie était un peu gris ce soir-là.

Sylvie's brother was a little drunk that evening.

2.13 Nous nous étions arrêtés quelques instants dans la maison du garde,—où, ce qui m'a frappé beaucoup, il y avait un cygne éployé sur la porte,

For a while we stopped at the keeper's house—where I was greatly struck to see a swan with spread wings displayed above the door,

2.14 puis au-dedans de hautes armoires en noyer sculpté, une grande horloge dans sa gaine, et des trophées d'arcs et de flèches d'honneur au-dessus d'une carte de tir rouge et verte.

and inside some tall cupboards of carved walnut, a large clock in its case, and trophies of bows and arrows of honor over a red and green target.

2.15 Un nain bizarre, coiffé d'un bonnet chinois, tenant d'une main une bouteille et de l'autre une bague, semblait inviter les tireurs à viser juste.

An odd dwarf, wearing a Chinese cap, and holding a bottle in one hand and a ring in the other, seemed to be inviting the marksmen to aim true.

2.16 Ce nain, je le crois bien, était en tôle découpée.

The dwarf, I am sure, was cut out of sheet-iron.

2.17 Mais l'apparition d'Adrienne est-elle aussi vraie que ces détails et que l'existence incontestable de l'abbaye de Châalis?	But is the apparition of Adrienne as real as these details, as real as the indisputable existence of the Abbey of Châalis?
2.18 Pourtant c'est bien le fils du garde qui nous avait introduits dans la salle où avait lieu la représentation;	Yet I am certain it was the keeper's son who took us into the hall where the play took place;
2.19 nous étions près de la porte, derrière une nombreuse compagnie assise et gravement émue.	we were near the door, behind a large audience, who were seated and seemed deeply moved.

Sentence (2.11) is an abstract stating the theme—dream versus reality—which the rest of the paragraph develops. From (2.12) to (2.17) the reality of Adrienne is questioned, and it is reasserted in (2.18) and (2.19), first by the bare statement of its reality (2.18), then by the description of undeniably real details (2.19). The segment from (2.12) to (2.17) itself breaks into two parts: from (2.12) to (2.16) details which are unexpected and thus undeniably real are described, and contrasted with the apparition of Adrienne in (2.17). Sentence (2.12) conveys a fact of coarse reality. But the reality of (2.13) and (2.14), describing the unusual swan and clock and embedded within the epistemic "ce qui m'a frappé beaucoup..." is somewhat attenuated. The description of the dwarf in (2.15) is positively dreamlike in its bizarreness, and in the failure to say whether he is real or artificial. Reality is re-established in (2.16) with the mention of sheet-iron, but even here the epistemic "je le crois bien" mitigates, paradoxically, the certainty it expresses.

The structure of this paragraph is illustrated in Figure 7.4.

7.2.4   The Return

In the next singing scene, N attempts to place the plain peasant girl Sylvie into the role of Adrienne—ironically, since N had loved Sylvie in childhood until the first moment he had seen Adrienne. This scene is remarkable for what is absent.

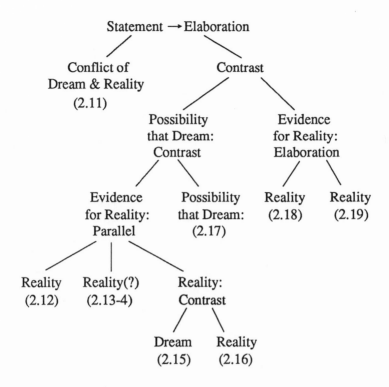

Figure 7.4 The Structure of (2.11)–(2.19).

3.1	Alors j'eus le malheur de raconter l'apparition de Châalis, restée dans mes souvenirs.	Then I was unlucky enough to tell her about the apparition at Châalis, which had remained in my memory.
3.2	Je menai Sylvie dans la salle même du château où j'avais entendu chanter Adrienne.	I took Sylvie to the very hall of the chateau where I had heard Adrienne sing.
3.3	— Oh! que je vous entende! lui dis-je;	"Oh, do let me hear you!" I said to her.
3.4	que votre voix chérie résonne sous ces voûtes	"Let your dear voice echo beneath these roofs
3.5	et en chasse l'esprit qui me tourmente, fût-il divin ou bien fatal! —	and drive away the spirit that torments me, whether it be from heaven or from hell!"
3.6	Elle répéta les paroles et le chant après moi:	She repeated the words and the song after me:

3.7 Anges, descendez promptement	Angels of Heaven, descend without delay
Au fond du purgatoire!...	To the pit of purgatory!...
3.8 — C'est bien triste! me dit-elle.	"It's very sad," she said.
3.9 — C'est sublime...	"It's sublime...
3.10 Je crois que c'est du Porpora, avec des vers traduits au XVI siècle.	"I think it's by Porpora, with words translated in the Sixteenth Century."
3.11 — Je ne sais pas, répondit Sylvie.	"I don't know," Sylvie answered.

N's intention in this episode is explicit. He wishes Sylvie's song to echo beneath the roofs as Adrienne's had, thereby driving out Adrienne's spirit and N's inexplicable confusion. The apparition of Adrienne will be replaced by the reality of Sylvie. The outcome is quite different however. Sylvie does not appear there magically as Adrienne had but has to be urged by N to take her place. She can only repeat the song, she cannot really sing it. We are told the subject matter by means of the bare reality of a direct quotation rather than by N's enraptured report as before. There is no mention at all of the quality of her voice or of whether she has trilled. Moreover, there is no light and no awed assembly.

In their discussion of the song afterwards, they talk past one another. Sylvie evaluates it in direct terms of the emotion produced, while N expresses an aesthetic judgment in terms appropriate to a sophisticated Parisian. N sinks to erudition. Sylvie responds with a statement that is beautifully ambiguous between an admission of ignorance and a rejection of the entire episode. N's effort to drive out the spirit of Adrienne with Sylvie has failed, and with it has failed his chances of loving Sylvie. On the return trip N decides Sylvie is no more than a sister to him and his thoughts turn to Aurélie.

7.2.5 Aurélie

In the final scene we analyze, N attempts to place Aurélie in the role of Adrienne, and there are even fewer descriptive details here than in Sylvie's singing scene.

4.1 J'avais projeté de conduire Aurélie au château, pres d'Orry, sur la même place verte où pour la prèmiere fois j'avais vu Adrienne.

I had planned to take her [Aurélie] to the chateau near Orry to the same square of green where for the first time I had seen Adrienne.

4.2 Nulle émotion ne parut en elle.

She showed no emotion.

4.3 Alors je lui racontai tout;

Then I told her everything;

4.4 je lui dis la source de cet amour entrevu dans les nuits, rêvé plus tard, réalisé en elle.

I told her the origin of that love half-seen in my nights, then dreamed of, then realized in her.

4.5 Elle m'écoutait sérieusement

She listened to me seriously

4.6 et me dit: — Vous ne m'aimez pas!

and told me: "It's not me you are in love with.

4.7 Vous attendez que je vous dise: La comédienne est la même que la religieuse;

You expect me to say, 'The actress is the same person as the nun.'

4.8 vous cherchez un drame, voilà tout,

You are simply seeking for drama, that's all,

4.9 et le dénouement vous échappe.

and the end eludes you.

4.10 Allez, je ne vous crois plus!

Go on, I don't believe in you any more."

In this episode, as in the previous one, N's intentions are clear. He has already decided that Aurélie fills the role of Adrienne in his life. Now he wants her to understand that, verifying its truth for himself. He wishes to tell her in the very place where his confusion began, and perhaps have her drive out Adrienne's spirit with her singing as Sylvie was to have done.

But the results are disastrous for N. The place is the same, but nothing else. There is no singing, no moonlight, no awe. There is only Aurélie's denial that she is Adrienne, and her refusal to occupy that role in N's life. Shortly after this incident Aurélie leaves N for another man. The romantic image of Adrienne, the inexplicable confusion, has destroyed his relationship with the other woman in his life.

7.3   The Temporal Structure and the Transitions

7.3.1   The Temporal Structure

This section is largely an explanation of Figure 7.5, which illustrates what we know about the time of the events in *Sylvie*, as well as how the story itself traverses these events. We first consider the part of the diagram that is in solid lines. A point, or node, in this diagram represents an event whose duration can be ignored for our present purposes. Each node represents the event that the label of the node indicates. An arrow from one node to another indicates that the time of the event represented by the first node precedes (or more correctly, does not follow) the time of the event represented by the second node. A double arrow between nodes means that their times coincide. Events that occur across some interval of time are represented by an arrow between two nodes, the nodes representing the beginning and end of the event. The diagram is not a straight line because there is much we do not know about the relative order of the events.

Thus, the earliest event (in absolute time) was Adrienne singing at the unnamed chateau. This was followed by Adrienne singing at Châalis at time T1b and also by the first visit to Loisy and Sylvie at time T1a, but the relative order of these two events is not known. Following both these episodes are the events at time T2. We have indicated in the figure only those events that are significant in the transitions in the story. Thus, N's leaving the theater is followed by his lying in bed, which is followed by his getting a coach, followed by his riding through the hills, followed by his riding past Orry, followed by his riding past Hallate, followed by the coach stopping, followed by the rest of the events of his second visit with Sylvie at Loisy. Then comes his return to Paris and his involvement with Aurélie during the period T3. During this period, he took Aurélie to the unnamed chateau and subsequently Aurélie's company performed at Dammartin. Following T3 is a period T4 during which N customarily visits Sylvie at Dammartin. The other event represented on the diagram is Adrienne's death in 1832. About this we know that it precedes the visit of Aurélie's company to the neighborhood of Dammartin. If the vision of Adrienne

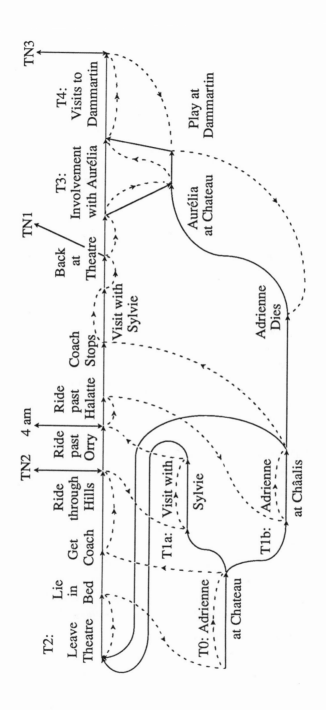

Figure 7.5 Temporal Structure of *Sylvie*.

at Châalis is a memory and not a dream, then we also know Adrienne's death follows time T1b. We know nothing else for certain. This may or may not mean Adrienne's death preceded time T2. But in Chapter 11, at time T2, there is the following passage:

Qu'est devenue la religieuse? dis–je tout à coup. —Ah! vous êtes terrible avec votre religieuse.....Eh bien! cela a mal tourné. Sylvie ne voulut pas m'en dire un mot de plus.	"What has become of the nun?" I suddenly asked. "You and your nun.....Well, you see, that had an unhappy ending," Sylvie would not tell me another word.

Also represented on the diagram is a trace (the dashed line) of how the story itself progresses through the events. The story begins as N leaves the theater and follows him to bed. It then switches to the events at time T0. It returns to T2 as N gets up and finds a coach, and follows him into the hills where it then jumps back to the events at time T1a. It returns to time T2 at four o'clock in the morning as N is riding past Orry. The coach continues and as it passes Hallate, the story takes us to the events at time T1b. Next, the story returns to time T2 as the coach stops and N disembarks. It then follows the real temporal order in a straightforward manner, first describing the visit with Sylvie at time T2, then the involvement with Aurélie during T3 (including Aurélie at the unnamed chateau, but not including the performance at Dammartin), and finally N's visits with Sylvie at Dammartin. In the last paragraphs of the book however, we are taken back to two previous events. First the narrator says that he forgot to mention a conversation with Sylvie during the performance at Dammartin in which Sylvie rejects the comparison of Adrienne and Aurélie. Then Sylvie tells N of Adrienne's death, taking us back in time to that event as the story closes.

Also represented on this diagram is the apparently shifting time of narration. Originally, the time of narration is a time we call TN1. The story begins

Je sortais d'un théâtre ou tous les soirs je paraissais.....	I came out of a theater where I used to spend money every evening.....

telling us that TN1 is after all the events at time T2, for the last thing he does at time T2 is return to Paris to go to the theater. At the end of Chapter 3, however, Nerval moves the story into the present tense and explicitly establishes the time of the telling as the time of riding through the hills. We call this time of narration TN2. In Chapter 8, the time of narration is reestablished at time TN1, and the past tenses are used for the events of time T2 again. The passages that effect this change in time of narration are examined below in Section 7.3.3.

Finally, in Chapter 14 the story again turns to the present tense, indicating that the events of the story have caught up with the time of the telling. We call this time of narration TN3. It may coincide with time TN1, but in Section 7.3.4 we argue that the story may have in fact surpassed the original time of narration. While time TN1 certainly does not follow time TN3, we do not know that it necessarily coincides with it.

Not represented in this diagram is an important aspect of how the story is organized—the epistemic status of the various episodes. The events in the past tend to be more dream-like and thus more representative of the romantic pole of the story's tension, whereas the more recent events have a greater sense of reality. The dream-like atmosphere is especially present in the two episodes with Adrienne. In Section 7.3.2, we examine how Nerval manipulates not only the time of the story but also its epistemic status.

While we are diagramming the relationships between order of events and order of telling, we may look closely at a particularly interesting case, by turning up the degree of magnification on the events at time T1b. This is illustrated in Figure 7.6. The events in order are the ride in Sylvie's brother's cart, the visit to the keeper's house at Châalis, going into the hall, hearing the other angels sing, and finally hearing Adrienne sing. The text covers these events in order, the last two being covered in what we presented above as (2.1) to (2.10). But then the same events, from the visit to the keeper's house, are described again, in more or less detail, in (2.11) to (2.19). As discussed in Section 7.2.3, this repetition highlights the question about the epistemic status of the episode—whether it is the memory of real events or only

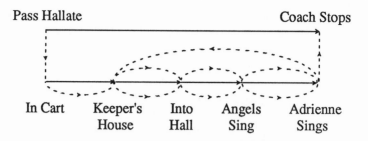

Figure 7.6 Temporal Structure of Chapter 7.

a dream. The episode thus exemplifies the interdependencies among the order of events, the order of their telling, and their epistemic status, in Nerval's story.

The next three sections of this chapter examine more closely the transitions Nerval uses to move between the different times in the story. There is a correlation between the times of the episodes and how they function in the tension between romance and reality. The earliest memories are the most romantic and the most dreamlike. As time progresses to T3 and T4, ordinary reality exerts an increasingly firm grip. Thus, the transitions between times should exhibit the essential tension between romance and reality in particularly striking ways, and should therefore repay microanalysis.

7.3.2   The Transition to the First Memory of Adrienne

The transition from the after-theater scene of Chapter 1 to the dreamlike memories of Adrienne in Chapter 2 occurs appropriately enough as N lies between wakefulness and sleep.

5.1	Je regagnai mon lit et je ne pus y trouver le repos.	I went to bed but could not rest.
5.2	Plongé dans un demi-somnolence,	Lost in a kind of half-sleep,
5.3	toute ma jeunesse repassait en mes souvenirs.	all my youth passed through my memory again.
5.4	Cet état, où l'esprit résiste encore aux bizarres combinaisons du songe,	This state, when the spirit still resists the strange combinations of dreams,

I	Remember	Youth
(5.3) mes	repassait en ... souvenirs	toute ma jeunesse
my	passed through ... memory	all my youth
(5.5) se	de voir ... presser en quelques minutes	les tableaux les plus saillants d'une longue période de la vie
us	to compress into a few moments	the most salient pictures of a long period of life
(5.6) je	me représentais	un château du temps de Henri IV
I	fancied myself	a chateau of the time of Henry IV

Figure 7.7 Parallelisms in the Transition.

5.5 permet souvent de voir se presser en quelques minutes les tableaux les plus saillants d'une longue période de la vie.

often allows us to compress into a few moments the most salient pictures of a long period of life.

5.6 Je me représentais un château du temps de Henri IV...

I fancied myself a chateau of the time of Henry IV...

Sentence (5.1) sets the scene by describing N's state, and clause (5.2) elaborates on that state. Then (5.3) through (5.6) elaborate the same theme in various ways. Clause (5.3) describes N's experience. Clauses (5.4) and (5.5) generalize this to a state everyone experiences, emphasizing it with a contrast. Sentence (5.6) begins the description of the specific contents of this particular experience, leading to the memory of the first encounter with Adrienne. But there are subtle differences that appear when we examine the structure of the elaborations closely, and these differences are significant.

Figure 7.7 shows schematically the deep parallelisms among the assertions of the text by lining up the similar items in columns. All assertions are instantiations of the general theme of remembering youth.

In assigning the memories, he begins with himself (5.3), generalizes to all people (5.5), then returns to himself (5.6). What is remembered progresses from the general "toute ma jeunesse" (5.3), to the slightly less general "les tableaux les plus saillants" (5.5), and finally to the specific chateau (5.6). The remembering itself is described first in terms that presuppose the reality of the memory—"repassait en mes souvenirs." The next expression of the remembering, "permet souvent de voir se presser en quelques minutes" is neutral with respect to the reality of what is seen, although the object of the predication does presuppose its own reality. This is contrasted with the spirit's resistance to dreams in (5.4), thus introducing the unreality of dreams as a possibility. Finally in (5.6) "me représentais" carries no presupposition of reality and in fact suggests unreality, leaving us open to the possibility that none of this actually occurred.

Thus has Nerval carried us from the reality of N's bed in Paris to the dreamlike events at the chateau in his childhood. By a succession of small changes embedded in a chain of elaborations he manipulates not only the time of events but also their epistemic status.

7.3.3   The Other Embedded Memories

We next examine a transition that is particularly interesting because it is just where the reader is likely to become confused about time. We try to pinpoint what the reader can miss that would lead to the confusion. The relevant passage spans a chapter break.

6.1   Quelle triste route, la nuit, que cette route de Flandre,

What a dreary track that Flanders road is at night.

6.2   qui ne devient belle qu'en atteignant la zone des forêts!

It only becomes beautiful when you reach the forest region.

6.3   Toujours ces deux files d'arbres monotones qui grimacent des formes vagues; au-delà, des carrés de verdure et de terres remuées, bornés à gauche par les collines bleuâtres de Montmorency, d'Écouen, de Luzarches.

All the time those two lines of monotonous trees, grimacing in vague shapes; beyond them square slabs of green, and of ploughed earth, bounded on the left by the bluish hills of Montmorency, Écouen, and Luzarches.

6.4 Voici Gonesse, le bourg vulgaire plein des souvenirs de la Ligue et de la Fronde...	Here is Gonesse, a vulgar little town full of memories of the Ligue and the Fronde.
6.5 Plus loin que Louvres est un chemin bordé de pommiers	Beyond Louvres is a road bordered by apple trees,
6.6 dont j'ai vu bien des fois les fleurs éclater dans la nuit comme des étoiles de la terre:	whose flowers I have often seen explode in the night like stars from the earth:
6.7 c'était le plus court pour gagner les hameaux.	It was the shortest way to reach the hamlets.
6.8 Pendant que la voiture monte les côtes,	While the carriage is climbing the sides of the hills,
6.9 recomposons les souvenirs du temps où j'y venais si souvent.	let's put in order the memories of the times when I came here so often.

IV. Un Voyage a Cythère	**IV. A Voyage to Cythera**
6.10 Quelques années s'étaient écoulées:	Some years had gone by.
6.11 l'époque où j'avais rencontré Adrienne devant le château n'était plus déjà qu'un souvenir d'enfance.	The time when I had met Adrienne in front of the chateau was already only a memory of childhood.
6.12 Je me retrouvai à Loisy au moment de la fête patronale.	I was at Loisy once again, at the time of the annual festival.
6.13 J'allais de nouveau me joindre aux chevaliers de l'arc, prenant place dans la compagnie dont j'avais fait partie déjà.	Once again I joined the knights of the bow and took my place in the company I had been part of before.

The first point of interest is the way in which Nerval moves us step by step from the past definite and imperfect tenses in which the last half of Chapter 3 has been told to the present tense in the final sentence of that chapter referring to a time that the narrator and reader share. The generalized present is used in (6.1)–(6.3) for a description that is always true. In (6.4) the deictic "voici" increases the immediacy of what is told by bringing the reader into the picture. The coach is now in the present and the reader is in the coach. In (6.5)–(6.7) the narrator begins to introduce the memories occasioned by the landscape, that he is about to expand upon, and the tense that is used could be used if the narrator were in the coach and the time of narration were the present. Finally in (6.8) and (6.9) the reader and the narrator are together in the coach at the time of narration which is the present, and the past tenses can now be used for events occurring before time T2.

In the first few sentences of Chapter 4 a confusion often arises. Sentence (6.10) is doubly indeterminate. Some years had gone by since when? This we learn from sentence (6.11) provided we assume it to be an elaboration of (6.10). But there is also an indeterminacy in the pluperfect tense of (6.10). Implicit in the pluperfect tense is some past point of reference prior to the "present." But it is uncertain what is meant by the "present." There are two possibilities, and they have statements in terms of the global structure of the story.

Chapters 1, 2 and 3 have all begun by describing successive events at time T2 and the events have been told primarily in the past tenses. A reasonable expectation is that Chapter 4 is at the same level in the global structure as the previous chapters, beginning as a further development of the events of the previous chapters. Moreover, it is reasonable to expect that the past point of reference implicit in the pluperfect tense of (6.10) refers to the time T2. That is, the reader can expect the "present" to be the original time of narration TN1, and the point of reference in the past to be time T2, the time of N's journey to Loisy, just as has been the case in the three previous chapters. It is a common mistake for the reader to assume just this until reaching Chapter 7, which forces a reinterpretation. The chapter break leads the reader to an incorrect structural analysis of the position of Chapter 4 and thus an incorrect assignment of its location in time. (By way of personal testimony, both authors of this analysis were victims of this confusion on their first readings.)

The other possibility is that Chapter 4 is not a continuation of Chapter 3, but an elaboration on (6.9), the last sentence in Chapter 3, and thus on the same structural level in the story as that sentence. (Figure 7.8 illustrates the two readings.)

If this is so, the "present" is the present time so carefully established by Nerval in the final sentences of Chapter 3 as time T2, and the events of Chapters 4, 5 and 6 take place at some

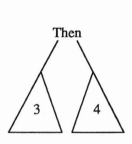

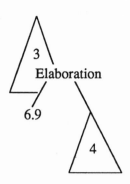

| **Figure7.8a** | **Figure7.8b** |
| Incorrect Structure. | Correct Structure. |

previous time, which we call time T1a. This turns out to be the correct interpretation.

The events of Chapters 4, 5 and 6 take place at time T1a and are told in the imperfect and past definite tenses. To return to time T2, Nerval simply switches back to the present tense with which he closed Chapter 3. He reinforces this return by telling us the time of night (6.14), and continuing the same sort of description of the passing landscape (6.15) with which he ended Chapter 3. He takes us a few more villages down the road (6.16), and by sentence (6.17) he has firmly reestablished the time as T2.

6.14 Il est quatre heures du matin; It is four in the morning;

6.15 la route plonge dans un pli de the road plunges into a dip of
terrain; elle remonte. land and then rises again.

6.16 La voiture va passer à Orry, The carriage is going by Orry,
puis à La Chapelle. then on to La Chapelle.

6.17 A gauche, il y a une route qui To the left there is a road that
longe le bois d'Hallate. runs along the wood of Hallate.

Now occasioned by a place he passes, the narrator launches into another memory, or almost a dream, since the epistemic status of Chapter 7 is the most uncertain in *Sylvie*. Its time T1b bears an uncertain relation to time T1a.

6.18 C'est par là qu'un soir It was along there that

6.19 le frère de Sylvie m'a conduit Sylvie's brother drove me one
dans sa carriole à une solennité evening in his little cart to a
du pays. country ceremony.

6.20 C'était, je crois, le soir de la Saint-Barthélemy.

It was, I believe, Saint Bartholomew's Eve.

6.21 A travers les bois, par des routes peu frayées, son petit cheval volait comme au sabbat.

His little horse flew through the woods and unfrequented roads as if to some witches' Sabbath.

6.22 Nous rattrapâmes le pavé à Mont-l'Evêque, et quelques minutes plus tard nous nous arrêtions à la maison du garde, à l'ancienne abbaye de Châalis.

We reached the paved road again at Mont l'Eveque and a few minutes later stopped at the keeper's lodge at the ancient Abbey of Châalis—

6.23 Châalis, encore un souvenir!

Châalis, yet another memory!

6.24 Cette vieille retraite des empereurs n'offre plus à l'admiration que les ruines de son cloître aux arcades byzantines, dont la dernière rangée se découpe encore sur les étangs, reste oublié des fondations pieuses comprises parmi ces domaines qu'on appelait autrefois les métairies de Charlemagne.

This former retreat of emperors now merely offers for our admiration the ruins of its cloisters with their Byzantine arcades, the last of which still stands out reflected in the pools—a forgotten fragment of those pious foundations included in the properties that used to be called "the forms of Charlemagne."

6.25 La religion, dans ce pays isolé du mouvement des routes et des villes, a conservé des traces particulières du long séjour qu'y ont fait les cardinaux de la maison d'Este à l'époque des Medicis.....

In this district, cut off from the movement of roads and cities, religion has preserved especial traces of the long stay made there by the Cardinals of the House of Este in the times of the Medici.....

6.26 Nous étions des intrus, le frère de Sylvie et moi, dans la fête particulière qui avait lieu cette nuit-là.

We were intruders, Sylvie's brother and I, in the private festival that took place that night.

This passage is an interesting example of the way in which shifts in tense produce an effect of confusion about the temporal sequence of the story. Not only is it impossible to anchor the time T1b at any specific point in the past, but a very curious relation is established between T1b and T2.

In (6.18) we are taken from T2 and the situation of the trip, to the past time T1b developed in (6.19), but the use of the present tense and of the deictic "là" somehow keeps us in the present. Moreover, the "c'est...que" construction anchors us in the present by presupposing the events it introduces. The use of the perfect tense "m'a conduit" in (6.19) maintains consistency with the present tense of (6.18) while moving the events into the past. By these means, sentence (6.18)–(6.19) introduces the remembered events of the past time T1b while linking them strongly to the present time T2. Sentence (6.20) specifies the time of the events of T1b, but in a way that increases rather than decreases the indeterminacy. We are not told the year nor any relationship with other episodes of the story, only the day of the year. Moreover, the phrase "je crois" calls even this specification into question.

Sentences (6.21) and (6.22) carry the narration forward, telling of the arrival at Châalis. Sentence (6.23) is an abstract characterization of the entire episode. It performs two functions: it frames the event by giving the category in which it must be interpreted, and anchors the temporal point of view again in T2. Sentence (6.23) cannot refer to any time but T2.

After this shift back to T2, we have in (6.24) and (6.25) a long descriptive digression which stops the narration of events. There is a problem in this description. The tense is the present tense, and the temporal adverbs "ne ...plus" and "encore" assume a temporal point of reference, which is the present. The problem is to decide whether the present is the time T2 or the time T1b, or a period that encompasses both. Is the deictic "encore" connected with the time of his actual trip or with the time of the memory? Since in the previous sentence (6.23), the present was T2, the most probable interpretation is that the present is still T2. But "ne...plus" in (6.24) gives us a second problem. At what point in the past was the abbey otherwise? It is reasonable to assume that the change described happened in an interval between T1b and T2, and the "ne...plus" negates what was true at T1b, since the "encore" refers to T2. During the passage (6.25), however, we realize that the interval is much longer and that the change has occurred since the time of the Medicis. This ambiguity depends on how large the present is assumed to be. If we assume a narrow present T2, the past is T1b, but if we assume a larger present including both T1b and T2, the past shifts back to a historical time, which turns out to be the right interpretation. But in this interpretation times T1b and T2 are, in a sense, both indistinguishably "present." By adopting a perspective in which times T1b and T2 do not differ significantly, the narrator has increased the indeterminacy of T1b, thereby contributing to the unreality of the episode.

Finally in (6.26) the narrator returns to the imperfect tense using the cohesion of "cette nuit-là" with elements in (6.18) and (6.20) to immerse us in the events of time T1b.

His memory, or dream, of Adrienne at Châalis ends abruptly at the end of Chapter 7.

6.27	Ce souvenir est une obsession peut-être!	Perhaps this memory is an obsession!
6.28	Heureusement voici la voiture qui s'arrête sur la route du Plessis;	Luckily the carriage stops here on the road to Plessis;
6.29	j'échappe au monde des rêveries,	I escape from the realm of reverie
6.30	et je n'ai plus qu'un quart d'heure de marche pour gagner Loisy par des routes bien peu frayées.	and have only a quarter of an hour's walk over little-used paths to reach Loisy.

VIII. Le Bal de Loisy   **VIII. The Ball at Loisy**

6.31	Je suis entré au bal de Loisy à cette heure mélancolique et douce encore où les lumières pâlissent et tremblent aux approches du jour.	I entered the ball at Loisy at that melancholy yet still gentle hour when the lights grow pale and tremble at the approach of day.
6.32	Les tilleuls, assombris par en bas, prenaient à leur cimes une teinte bleuâtre.	The lime trees, in deep shadows at their roots, took on a bluish tint at the top.
6.33	La flûte champêtre ne luttait plus si vivement avec les trilles du rossignol.	The bucolic flute no longer struggled so keenly with the song of the nightingale.
6.34	Tout le monde était pâle, et dans les groupes dégarnis j'eus peine à rencontrer des figures connues.	Everyone looked pale and in dishevelled groups I had difficulty finding faces I knew.

6.35 Enfin j'aperçus la grande Lise, une amie de Sylvie.	At last I saw Lise, a friend of Sylvie.
6.36 Elle m'embrassa.	She kissed me.

Nerval carries time T2 back to the past gradually. First in (6.31) he uses the perfect tense, just as at the beginning of Chapter 7, he used the perfect tense to mediate between the present and the past. Here it moves us forward slightly. In (6.30) N is on the road. In (6.31) it must be at least fifteen minutes later, and in fact may be quite a bit later. The next two sentences, (6.32) and (6.33), are in the imperfect tense, which probably means that we have been returned to the original time of narration TN1, but they are general descriptive passages. Then in the next three sentences (6.34)–(6.36) he moves us at last into particular events told in the past definite tense, firmly bringing the time of narration back to TN1.

Nerval faces two problems in changing from the present in (6.30) to the perfect in (6.31). Since a chapter break intervenes, the reader is much freer in the structural connection he interprets (6.31) as having. Nerval must get the reader to see it as a continuation of (6.30). He faces the further problem of getting the reader to draw the implicature that fifteen minutes have passed at time of narration TN2, a rather unusual thing to have happened during a narration. He overcomes these problems by using the strong occasion relation between "un quart d'heure de marche pour gagner Loisy" in clause (6.30) and "je suis entré au bal de Loisy" in sentence (6.31). That is, entering the ball at Loisy is a reasonable thing to happen after a walk to Loisy. Moreover, he elaborates on the time of the night, which is an unusual time to be out and about and was prominent in every other transition between time T2 and previous events. The strong coherence of event and clock time thus establish the link between (6.30) and (6.31) firmly enough that Nerval is able switch tenses in (6.31). This makes possible the subsequent development of events beyond time T2, as Nerval needs for the rest of the story.

7.3.4 Time Speeds Up

In Chapter 13 something strange happens. Prior to this, the events at time T2 have proceeded slowly. Nerval has established a temporal framework in the early chapters of the story which is anchored in T2. The events at time T2 are told with great attention to detail and other events are told as memories from time T2. It would be reasonable to expect the story to be brought to a conclusion at time T2. But suddenly at the end of the first paragraph of Chapter 13, with sentence (7.4), time speeds up.

7.1   Pendant le quatrième acte, où elle ne paraissait pas, j'allai acheter un bouquet chez madame Prévost.

During the fourth act, when she did not come on, I went and bought a bouquet of flowers at Madame Prevost's.

7.2   J'y insérai une lettre fort tendre signée: *Un inconnu.*

In it I placed a most tender letter signed "An Unknown."

7.3   Je me dis: Voilà quelque chose de fixé pour l'avenir,

I said to myself, That's something of the future settled.

7.4   —et le lendemain j'étais sur la route d'Allemagne.

And the next day I was travelling to Germany.

Rather than remain at time T2, the story moves on rapidly, first through a period of some months during which N becomes involved with Aurélie, a period we have called T3. Finally, in Chapter 14 Nerval brings us up to habitual events told in the present tense, at a time we call T4. Time has advanced enough for Sylvie to marry and have two children.

This sudden change in the "grain" of the story somehow gives us the feeling that the time of the story has caught up with and even passed the time of the telling. The time of narration TN3 seems later than time TN1.

One factor contributing to this feeling involves the explicit identification of the time of events with the time of narration at two places in the story. The first is at the beginning of Chapter 3 and again at the end in the sentences analyzed above as (6.4) and (6.8)–(6.9). There the time of narration was identified with a time in the middle of the period T2. Then at the beginning of Chapter 14, we find the following:

7.5   Telles sont les chimères qui charment et égarent au matin de la vie.

Such are the delusions which charm and beguile us in the morning of life.

7.6 J'ai essayé de les fixer sans beaucoup d'ordre, mais bien des cœurs me comprendront.

I have tried to set them down without too much order but many hearts will understand mine.

The remainder of Chapter 14 continues in the habitual present, placing time TN3 at a point several years after time TN2. All this does not place time TN3 definitively after the original time of narration TN1, but at the very least, it tells us that the time of narration is fluid.

Moreover, the geography which plays such an important role in Chapters 3 through 8 in organizing the shifts in time carries a romantic aura with it, appropriate for its occasioning of the romantic memories. But in Chapter 14, all this is denied.

7.7 Othys, Montagny, Loisy, pauvres hameaux voisins, Châalis,—que l'on restaure,—vous n'avez rien gardé de tout ce passé!

Othys, Montagny, Loisy, poor neighboring villages, Châalis (oh, that they would restore it), you have retained nothing of the past!

The geography that organizes the story at times of narration TN1 and TN2 would no longer seem capable of that function at time TN3.

7.4 Conclusion

We have analyzed four episodes central to the definition of story content, as well as the temporal structure and transitions which constitute an important feature of the textual organization of *Sylvie*.

These two different levels of analysis were chosen because they represent two different developments of the basic theme of *Sylvie*, on both the level of the story and the metalevel of the narration—N's failure to make sense of his life experience.

Each of the four episodes analyzed in Part 2 concerns N's attempt, and failure, to resolve the opposition between a romantic image and reality. We have shown how these episodes, all repetitions of the same scene, mark a progression in decreasing structural complexity, parallel to an increase in reality. The first episode of Adrienne introduces in the strongest way the theme of romantic and ideal love. It is elaborated in the second episode at a more abstract level. The ambiguity of the epistemic status of the first two episodes and their indeterminacy in time emphasize the theme of the romantic by cutting them loose from the world of reality. This is particularly clear in the sequence at Châalis in which Adrienne becomes the romantic image in a more symbolic and intangible way. The next two episodes describe N's attempts to fit reality into his romantic schema. He first tries with Sylvie, reproducing the scene of the song, and then with Aurélie, by asking her to recognize an identity between the nun and the actress. In both cases he fails. The two women will not go along, and the two worlds of dream and reality cannot be united. This failure is double: both dream and reality are lost. Adrienne, the ideal for whom N was searching, was already dead during much of his search. But reality has been lost as well. Sylvie marries his foster brother, and Aurélie comes to love (or better, be loved by) another man. This is not surprising: since N looks at reality only to interpret it according to his romantic schema, he is not able to see reality in itself, and so he loses it. The three women are purely functions for him. They instantiate abstract roles rather than having a concrete reality. They are therefore interchangeable, all the more remarkably since the three women, from the little we are told, differ so radically.

By trying to reduce reality to the romantic ideal, N experiences an existential failure, at the level of the story. This is reflected at the level of the telling of the story, where he experiences a cognitive failure, a failure to understand his past. The strong parallelism between past and present on the one hand and between romance and reality on the other is stated quite explicitly at the beginning of the last chapter:

8.1   Telles sont les chimères qui charment et égarent au matin de la vie.... .   Such are the delusions that charm and beguile us in the morning of life.... .

8.2   Les illusions tombent l'une après l'autre, comme les écorces d'un fruit, et le fruit, c'est l'expérience.   Illusions fall, like the husks of fruit, one after another, and what is left is experience.

In Chapter 3 N summarizes what he is trying to do with the story by saying,

8.3   recomposons les souvenirs du   let's put in order the memories
      temps ou j'y venais si souvent.   of the times when I came here
                                          so often.

Similarly in Chapter 13, N says,

8.4   Qu'allais-je y faire?   What was I going to do there?
8.5   Essayer de remettre de l'ordre   Try and get my feelings into
      dans mes sentiments.   order.

The narration itself is his attempt to put order into his feelings
and his memories, into his life experience.

What then is the significance of the vagaries of the temporal
structure? The construction of a life story requires at least that
we impose a temporal order on events. But in *Sylvie*, there is
a nonlinear organization of events. Transitions are not clear,
and subtle shifts of tense abound. The effect for the reader is
confusion, duplicating N's confusion of past and present, romance
and reality.

Among the most confusing elements for the reader of *Sylvie*,
and at the same time among the most peculiar features of the
writing, are the shifts and ambiguities in the time of narration.
The time of narration is the privileged point of view from which
the author looks at the events of the story to impose an order
upon them. But the shifts and ambiguities undermine the priv-
ilege. The multiplication of textual points of view derives from
the impossibility for N to put himself in a position from which
to reconstruct his experience, in a way that will allow his past
to function in his present.

In the last chapter N tries a kind of synthesis, a sober wisdom,
but he is not convincing. He tells us of experience that

8.6   Sa saveur est amère; elle a   It has a bitter taste, but
      pourtant quelque chose d'âcre   there is something tonic in its
      qui fortifie.   sharpness.

But as the chapter elaborates this theme, N vacillates between
the bitter and the tonic with the bitter more often than not
prevailing. Most telling are the last few sentences in the story.
N says that he forgot to mention something before, and then we
are told that the similarity of Adrienne and Aurélie is possibly
illusory. Finally in the last sentence of the story we are informed

that Adrienne is dead. Before the last sentence we could still interpret the last chapter as concluding the story with the lesson of "acquired experience," but after it this interpretation is no longer possible. The last sentence forces a reinterpretation of what N has told us about experience. This is a fact of the utmost importance and if N had indeed achieved a synthesis, this fact would be integrated into his account of the synthesis and not be told as an afterthought. We have to conclude that N has failed in his effort to achieve by means of this narration a synthesis of his dream-like past and the realities of the present.

Afterword

Jorge Luis Borges has a piece entitled, "An Examination of the Work of Herbert Quain," in which he says of Quain, "He thought that good literature was common enough, that there is scarce a dialogue in the street which does not achieve it." Microanalysis of linguistic material at any level reveals previously unsuspected complexity. When it is discourse we are microanalyzing, we are tempted to call this complexity artistry. Ordinary conversation is already a magnificent achievement, and literature is only a second-order effect on top of that.

Consider an example. It comes from a series of ethnographic life history interviews that Michael Agar conducted with a 60-year-old heroin addict, and that Agar and I have analyzed (Agar and Hobbs, 1982). We originally chose this fragment because nothing much is happening in it. The subject, Jack, has just told a good story about how he stole someone's luggage and he is just about to tell a good story about how he fenced the goods he found in the luggage. But this fragment is just connective tissue between the two.

(1) J: So I split up the street,
(2)   now remember, snow and ice,
(3)   I split up the street,
(4)   and at that time there used to be a Chase's cafeteria,
(5)   I don't know what it's called now,

(6)   but you know where the Selwyn Theatre is on 42nd
         Street?

         :

(7)   You know where Grant's is,
    M: Yeah.
(8)   J: you've heard of Grant's,
    M: Oh yeah.
(9)   J: Well just about three doors down from Grant's,
(10)   Chase's cafeteria.
(11)   It was open all night long,
(12)   and strictly a hangout after certain hours for hustlers.
    M: Uh huh.
(13) J: Across the street midway down the block was Bickford's,
(14)   I guess it's even still there,
(15)   maybe it isn't, I don't know,
(16)   but at any rate there was a Bickford's.
(17)   That was another hangout.
(18)   Then on— going back to the other side of the street,
         down—
(19)   you know where there— there's an arcade, a flea circus,
         an arcade?
(20)   Well that used to be a *bus* station at one time,
(21)   and you could go through there all the way to 41st
         Street.
(22)   And there were pinball games and all sorts of you know
         amusements,
(23)   and of course lots of hustlers hung out in *there* too.
(24)   And right next door to it was a Horn and Hardart's,
(25)   and of course you could go in there
(26)   for a nickel cup of coffee you could sit for hours.
(27)   Well I went to Horn and Hardart's that morning.

Jack has several goals at this point in the conversation. He
needs to get himself from the train station where he stole the
luggage to a cafeteria where he fenced the goods. He wants to
provide Agar with ethnographic information about New York
street life in the 1940s. And he wants to relate this to what Agar
knows of New York in the 1970s. What he produced turns out
to have quite an elegant structure, illustrated in Figure 8.1.

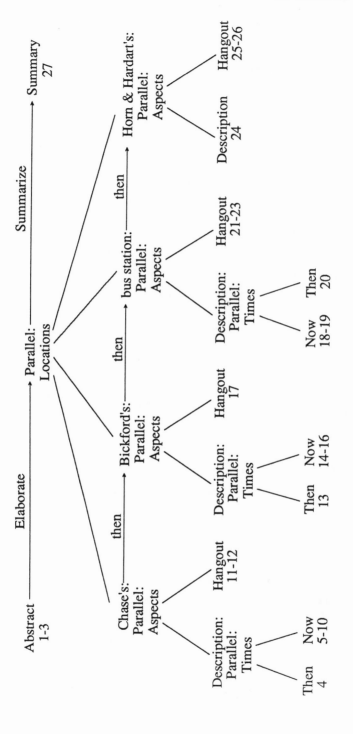

Figure 8.1 Structure of (1)-(27).

He takes us down the street in temporal order, describing the places he would have passed. But the places are not just anyplace. They are all hangouts, places where he might find a fence. For each of these places he gives a description and then says it was a hangout. All his descriptions but the last are given in two parts. First he describes what was there in the 1940s. Then to identify the place in terms of what Agar knows, he says what is there now.

Microanalysis frequently turns up this sort of artistry. But something else shows up under microanalysis that is just the opposite—namely, all the small-scale failures that beset our conversations, all the miscommunications. Frequently we go away from a conversation with the sense that it was successful when in fact we have just talked past each other. Evans and I (Hobbs and Evans 1980) microanalyzed a fragment of a videotaped conversation between a man and a woman, and then we interviewed the woman. When we showed her the tape and said we wanted to ask her some questions about it, her reaction was to say, "It seems very clear. What are your questions about?" Yet in our microanalysis we had amassed lots of evidence that the goals of the two participants were disastrously at odds with each other. The woman's aim in the conversation is to talk about her dissertation, which she just finished. The man's aim is to avoid embarrassing himself in front of the camera. The woman prominently displays her dissertation, which she is carrying in her arms along with a bundle of envelopes. The man asks about the envelopes. After describing these, the woman explicitly introduces the dissertation, but the man diverts the talk from its content to the question of whether he is cited.

Gumperz (1982) has analyzed striking examples of such failed conversations, and Tannen (1979) has turned up examples at dinner parties where you would least expect to find them. A favorite example of *mine* is a conversation between a radio talk show host who has asked people to call in and tell about their worst blind date, and a woman who calls in. So far he has had nothing but boring stories and he clearly expects nothing from this woman. So for the first half of the conversation, he is trying to make slightly risqué jokes at her expense, while she's trying to get on with her story. Suddenly she tells the host that she stole the car.

The host says, "You mean—wait a minute—you drove off. . . "
The woman says, "I left them way up there."
"You drove off in their car?"
"Yeh, I sure did."

From then on the host tries to get more good material out of her, but she has already told her story and only repeats it again.

This sort of failed conversation is hardly rare. In fact it is probably typical. But even if this is the case, even if most conversation is unsuccessful, successful conversations do occur and merit investigation, if for no other reason than that they are the ideal toward which all conversations tend. They are the reason that we engage in conversation at all. So we are led to ask what happens when conversations work, when communication occurs.

There are at least three cases in which conversations can succeed spectacularly. (There are no doubt many more.) The first case is when someone says something to us to provide the missing piece in the solution to a problem that we have been working on for some time, or phrases something in just the right way to give us the correct perspective on an issue. In this case, there is no need at all for the speaker to intend to do this for us, or even to be aware that it has happened. In fact, most of the time when this happens, we are unable to explain to the other what he or she has done for us. It is often a remark we will remember the rest of our lives and they will not recall tomorrow. A personal example of this is when I heard Michel Foucault referred to as someone who was "nostalgic for the good old days when the mad ran free." I have read Foucault in a different way ever since.

This is a case where conversations pay off for us as listeners. A second case is when the listener demands the best in us as speakers. They do not let us get away with less than we are capable of. An unexpected instance of this occurs to me sometimes when I'm traveling in some underdeveloped country and I start bargaining with somebody for something. The guy turns out to be particularly obdurate, and I have time to kill, so we spend half an hour haggling over whether I should spend $4 or $5. Well, you can't talk numbers for half an hour, so you're driven to more and more creative arguments, bordering on the whimsical and fantastical. For example, bargaining for a camel ride once while a sandstorm raged outside, I did a back-of-the-envelope calculation of the annual cost of maintaining a camel. The response was a detailed description of all the diseases camels are subject to. Sometimes one of you comes up with something that sounds so good that you win. But by that time it no longer matters who wins. The conversation has succeeded.

The third case is when we recognize just how much that matters to us is shared with the other. The conversation begins to play off this coincidence of beliefs and concerns. There is a certain irony here. The stereotypic view of language is that we use the information we share in order to convey new information. Frequently in this third type of conversation, the picture is turned upside down. We convey new information as an excuse to demonstrate the presuppositions we share with each other. Depth and extent in what is shared allows another kind of conversation as well, in which inferences and implicatures are possible that are not ordinarily possible. It allows us that joyful experience of being obscure and being understood nevertheless—even more, knowing we will be understood. The joy is not so much in solving the problem as in knowing we have the resources to solve the problem.

Let me give a personal example of this as well. All my life I was a terrible dancer until one August, I suddenly became a good dancer. The change was the result of an insight—namely, that dancing is a kind of discourse analysis.

I told this to a friend of mine, and she said, "Of course."

And then to prove that she was not just faking it, she told me about the waltz, in which you whirl around so much that the only way to keep from losing your balance is to keep your eyes on the *one fixed point* in your environment—your partner. This was precisely the right response to make, for what I had meant by my obscure remark was that dancing is not a matter of moving one's body the right ways, but rather is a matter of playing with the spatial relationship between one's self and one's partner in the context of the entire dance floor.

Now instead of considering the force of this example, let us consider its content. Paul Valery has characterized the difference between ordinary discourse and poetry by the following analogy: ordinary discourse is like walking from one place to another; a poem is like a dance. So let's suppose my insight about dancing is in fact true. Then what distinguishes poetic discourse is not so much the shape of the work that the writer executes. Rather, it is the special relationship he establishes with his reader, demanding the best of both writer and reader, communicating important insights, and demonstrating the depth to which we are understood.

The same is true of the best of conversation.

Bibliography

Agar, M., and J. R. Hobbs. 1982. Interpreting Discourse: Coherence and the Analysis of Ethnographic Interviews. *Discourse Processes* 5(1):1–32.

Allen, J. F., and R. Perrault. 1980. Analyzing Intention in Utterances. *Artificial Intelligence* 15:143–178.

Aristotle. 1975. *"Art" of Rhetoric.* Vol. 22 of *Loeb Classical Library.* Cambridge, MA: Harvard University Press. Translated by J. Freese.

Artin, E. 1959. Galois theory. Notre Dame Mathematical Lectures 2, Notre Dame, Indiana.

Bear, J., and J. R. Hobbs. 1989. Localizing Expression of Ambiguity. In *Proceedings of the Second ACL Conference on Applied Natural Language Processing*, 235–242, Austin, TX.

Beardsley, M. 1958. *Aesthetics: Problems in the philosophy of criticism.* New York: Harcourt Brace and Co.

Beardsley, M. 1967. Metaphor. In P. Edwards (Ed.), *Encyclopedia of Philosophy*, Vol. 5, 284–289. New York: Collier–Macmillan.

Beardsley, M. C., and W. K. Wimsatt Jr. 1954. The Interpretation Fallacy. In J. W. K. Wimsatt (Ed.), *The Verbal Icon: Studies in the Meaning of Poetry.* Lexington, Kentucky: University of Kentucky Press.

Bever, T. G. 1986. The Aesthetic Basis for Cognitive Structures. In M. Brand and R. M. Harnish (Eds.), *The Representation of Knowledge and Belief.* Tucson: University of Arizona Press.

173

Black, M. 1962. Metaphor. In M. Black (Ed.), *Models and Meta-phors: Studies in Language and Philosophy*, 25–47. Ithaca, New York: Cornell University Press.

Black, M. 1979. More about Metaphor. In A. Ortony (Ed.), *Metaphor and Thought*, 19–43. Cambridge, England: Cambridge University Press.

Brooks, C. 1965. Metaphor, Paradox, and Stereotype. *British Journal of Aesthetics* 5:315–328.

Bruce, B. C., and D. Newman. 1978. Interacting Plans. *Cognitive Science* 2:195–233.

Carbonell, J. 1982. Metaphor: An Inescapable Phenomenon in Natural-Language Comprehension. In W. Lehnert and M. Ringle (Eds.), *Strategies for Natural Language Processing*, 415–434. Hillsdale, New Jersey: Lawrence Erlbaum Associates.

Chafe, W. 1980. The Deployment of Consciousness in the Production of a Narrative. In W. Chafe (Ed.), *The pear stories: Cognitive and Linguistic Aspects of Narrative Production*. Norwood, New Jersey: Ablex Publishing Company.

Charniak, E., and R. Goldman. 1988. A Logic for Semantic Interpretation. In *Proceedings, 26th Annual Meeting of the Association for Computational Linguistics*, 87–94, Buffalo, New York. June 1988.

Clark, H. 1973. Space, Time, Semantics, and the Child. In T. Moore (Ed.), *Cognitive Development and the Acquisition of Language*. New York: Academic Press.

Collins, A., and M. R. Quillian. 1971. How to Make a Language User. In Tulving and Donaldson (Eds.), *Organization of memory*, 309–351. New York: Academic Press.

Crothers, E. 1979. *Paragraph Structure Inference*. Norwood, New Jersey: Ablex Publishing Corporation.

Dahlgren, K. 1985. The Cognition Structure of Social Categories. *Cognitive Science* 9(3):379–398.

Davidson, D. 1967. The Logical Form of Action Sentences. In N. Rescher (Ed.), *The Logic of Decision and Action*, 81–95. Pittsburgh, Pennsylvania: University of Pittsburgh Press.

Donnellan, K. S. 1966. Reference and Definite Descriptions. *The Philosophical Review* 75:281–304.

DuBois, J. W. 1987. Meaning without Intention: Lessons from Divination. *Papers in Pragmatics* 1(2).

Eliot, T. S. 1920. Tradition and the Individual Talent. In *The Sacred Wood*. London: Metheun and Company.

Evans, T. 1968. A Program for the Solution of a Class of Geometric-Analogy Intelligence-Test Questions. In M. Minsky (Ed.), *Semantic Information Processing*, 271–353. Cambridge, MA: The MIT Press.

Fillmore, C. 1974. Pragmatics and the Description of Discourse. In C. Fillmore et al. (Eds.), *Berkeley Studies in Syntax and Semantics*, Vol. 1, V–1–V–21. Berkeley, CA: University of California.

Fillmore, C. 1979. Innocence: A Second Idealization for Linguistics. In *Proceedings, Fifth Annual Meeting, Berkeley Linguistics Society*, 63–76, Berkeley, CA.

Fish, S. 1980. *Is There a Text in This Class?* Cambridge, MA: Harvard University Press.

Gentner, D. 1983. Structure-Mapping: A Theoretical Framework for Analogy. *Cognitive Science* 7:150–170.

Ginsberg, M. L. (Ed.). 1987. *Readings in Nonmonotonic Reasoning*. Los Altos, CA: Morgan Kaufmann Publishers, Inc.

Grice, H. P. 1975. Logic and Conversation. In P. Cole and J. Morgan (Eds.), *Syntax and Semantics*, Vol. 3, 41–58. New York: Academic Press.

Grimes, J. 1975. *The Thread of Discourse*. The Hague, Netherlands: Mouton and Company.

Grosz, B. 1977. The Representation and Use of Focus in Dialogue Understanding. Stanford Research Institute Technical Note 151, Stanford Research Institute, Menlo Park, CA, July.

Gumperz, J. 1982. *Discourse Strategies*. Cambridge, England: Cambridge University Press.

Hirsch, E. D. 1967. Objective Interpretation. In *Validity in Interpretation*. New Haven: Yale University Press. Reprinted from PMLA, vol. 75, September 1960.

Hirsch, E. D. 1976. *The Aims of Interpretation*. Chicago: University of Chicago Press.

Hobbs, J. R. 1976. A Computational Approach to Discourse Analysis. Research report 76-2, Department of Computer Sciences, City College, City University of New York, December.

Hobbs, J. R. 1977. From 'Well-Written' Algorithm Descriptions into Code. Research Report 77-1, Department of Computer Sciences, City College, City University of New York, July.

Hobbs, J. R. 1979. Coherence and Coreference. *Cognitive Science* 3(1):67–90.

Hobbs, J. R. 1980. Selective Inferencing. In *Proceedings of the Third National Conference of the Canadian Society for Computational Studies of Intelligence*, 101–114, Victoria, British Columbia, May.

Hobbs, J. R. 1982. Representing Ambiguity. In *Proceedings of the First West Coast Conference on Formal Linguistics*, 15–28, Stanford, CA, January.

Hobbs, J. R. 1985. Ontological Promiscuity. In *Proceedings, 23rd Annual Meeting of the Association for Computational Linguistics*, 61–69, Chicago, IL, June.

Hobbs, J. R. 1990. Topic Drift. In B. Dorval (Ed.), *Conversational Coherence and its Development*. Norwood, NJ: Ablex Publishing Corp.

Hobbs, J. R., and M. Agar. 1985. The Coherence of Incoherent Discourse. *Journal of Language and Social Psychology* 4(3-4):213–232.

Hobbs, J. R., W. Croft, T. Davies, D. Edwards, and K. Laws. 1987. Commonsense Metaphysics and Lexical Semantics. *Computational Linguistics* 13(3-4):241–250.

Hobbs, J. R., and D. A. Evans. 1980. Conversation as Planned Behavior. *Cognitive Science* 4(4):349–377.

Hobbs, J. R., and R. C. Moore (Eds.). 1985. *Formal Theories of the Commonsense World*. Norwood, New Jersey: Ablex Publishing Corporation.

Hovy, E. H. 1988. Planning Coherent Multisentential Text. In *Proceedings of the 26th Annual Meeting, Association for Computational Linguistics*, 163–169, Buffalo, New York, June.

Indurkhya, B. 1986. Constrained Semantic Transference: A Formal Theory of Metaphors. *Synthese* 68:515–551.

Indurkhya, B. 1987. Approximate Semantic Transference: A Computational Theory of Metaphors and Analogies. *Cognitive Science* 11(4):445–480. October-December 1987.

Isenberg, A. 1963. On Defining Metaphor. *The Journal of Philosophy* 60:609–622.

Jackendoff, R. 1976. Toward an Explanatory Semantic Representation. *Linguistic Inquiry* 7(1):89–150. Winter 1976.

Jespersen, O. 1922. *Language: Its Nature, Development and Origin*. London: George Allen and Unwin Ltd.

Kahn, E. 1975. Frame Semantics for Motion Verbs with Application to Metaphor. In C. Cogen et al. (Eds.), *Proceedings of the First Annual Meeting of the Berkeley Linguistic Society*, 246–256.

Kling, R. 1971. A Paradigm for Reasoning by Analogy. *Artificial Intelligence* 2:147–178.

Knapp, S., and W. B. Michaels. 1982. Against Theory. *Critical Inquiry* 8(4):723–42.

Knapp, S., and W. B. Michaels. 1987. Against Theory 2: Hermeneutics and Deconstruction. *Critical Inquiry* 14(1):49–68.

Knuth, D. 1973. *The Art of Computer Programming*. Vol. 1. Reading, MA: Addison-Wesley.

Lakatos, I. 1970. Falsification and the Methodology of Scientific Research Programmes. In I. Lakatos and A. Musgrave (Eds.), *Criticism and the Growth of Knowledge*. Cambridge.

Lakoff, G., and M. Johnson. 1980. *Metaphors We Live By*. Chicago: University of Chicago Press.

Langer, S. 1942. *Philosophy in a New Key: A Study in the Symbolism of Reason, Rite, and Art*. Cambridge, MA: Harvard University Press.

Lenat, D., M. Prakash, and M. Shepherd. 1986. CYC: Using Common Sense Knowledge to Overcome Brittleness and Knowledge Acquisition Bottlenecks. *AI Magazine* 6(4):65–85.

Levin, S. R. 1977. *The Semantics of Metaphor*. Baltimore: The Johns Hopkins University Press.

Lewis, D. 1979. Scorekeeping in a Language Game. *Journal of Philosophical Logic* 6:339–59.

Linde, C. 1990. *The Creation of Coherence in Life Stories*. In preparation.

Linde, C., and J. Goguen. 1978. Structure of Planning Discourse. *Journal of Social and Biological Structures* 1:219–251.

Longacre, R. 1976. *An Anatomy of Speech Notions*. Ghent, Belguim: The Peter de Ridder Press.

MacLane, S. 1971. *Categories for the Working Mathematician*. New York: Springer-Verlag.

Mailloux, S. 1985. Rhetorical Hermeneutics. *Critical Inquiry* 11:620–41.

Mann, W., J. Moore, and J. Levin. 1977. A Comprehension Model for Human Dialogue. In *Proceedings, International Joint Conference on Artificial Intelligence*, 77–87, Cambridge, MA. August 1977.

Mann, W., and S. Thompson. 1986. Relational Propositions in Discourse. *Discourse Processes* 9(1):57–90.

Matthews, R. 1971. Concerning a 'Linguistic Theory' of Metaphor. *Foundations of Language* 7:413–425.

McDermott, D., and J. Doyle. 1980. Non-monotonic Logic I. *Artificial Intelligence* 13(1,2):41–72.

McDermott, J. 1979. Learning to Use Analogies. In *Proceedings, International Joint Conference on Artificial Intelligence*, 568–576, Tokyo, Japan, August.

Meyer, B. 1975. Identification of the Structure of Prose and its Implications for the Study of Reading and Memory. *Journal of Reading Behavior* VII(1):7–47.

Miller, G. 1979. Images and Models, Similes and Metaphors. In A. Ortony (Ed.), *Metaphor and Thought*, 202–250. Cambridge, England: Cambridge University Press.

Montague, R. 1974. The Proper Treatment of Quantification in Ordinary English. In R. Thomason (Ed.), *Formal Philosophy: Selected Papers*. New Haven, Connecticut: Yale University Press.

Morgan, J. 1979. Observations on the Pragmatics of Metaphor. In A. Ortony (Ed.), *Metaphor and Thought*, 136–147. Cambridge, England: Cambridge University Press.

Nerval, G. de. 1957. *Selected Writings*. New York: Grove Press. Translated by Geoffrey Wagner.

Nerval, G. de. 1965. *Sylvie, Les Filles du Feu, Les Chimères*. Paris: Garnier-Flamarion.

Nunberg, G. 1978. *The Pragmatics of Reference*. PhD thesis, City University of New York, New York.

Ogden, C. 1932. *Bentham's theory of fictions*. London: Routledge and Kegan Paul Ltd.

Ong, W. 1955. Metaphor and the Twinned Vision. *Sewanee Review* 63:193–201.

Ortony, A. 1979. The Role of Similarity in Similes and Metaphors. In A. Ortony (Ed.), *Metaphor and Thought*, 186–201. Cambridge, England: Cambridge University Press.

Ortony, A., R. Reynolds, and J. Arter. 1978. Metaphor: Theoretical and Empirical Research. *Psychological Bulletin* 85(5):919–943.

Pollack, M. E. 1986. A Model of Plan Inference that Distinguishes between the Beliefs of Actors and Observers. In *Proceedings of the*

24th Annual Meeting of the Association for Computational Linguistics, 207–214, New York.

Reddy, M. 1979. The Conduit Metaphor—A Case of Frame Conflict in our Language about Language. In A. Ortony (Ed.), *Metaphor and Thought*, 284–324. Cambridge, England: Cambridge University Press.

Richards, I. A. 1936. *The Philosophy of Rhetoric*. Oxford: Oxford University Press.

Rieger, C. 1978. Viewing Parsing as Word Sense Discrimination. In W. Dingwall (Ed.), *A Survey of Linguistic Science*. Greylock Pub.

Rumelhart, D. 1979. Some Problems with the Notion of Literal Meanings. In A. Ortony (Ed.), *Metaphor and Thought*, 78–90. Cambridge, England: Cambridge University Press.

Russell, S. W. 1976. Computer Understanding of Metaphorically Used Verbs. *American Journal of Computational Linguistics*. microfiche 44.

Schank, R., and R. Abelson. 1977. *Scripts, Plans, Goals, and Understanding*. Hillsdale, New Jersey: Lawrence Erlbaum Associates. Inc.

Schiffer, S. R. 1972. *Meaning*. Oxford: Oxford University Press.

Schourup, L. C. 1985. *Common Discourse Particles in English Discourse*. New York: Garland Publishing Company.

Searle, J. 1979. Metaphor. In A. Ortony (Ed.), *Metaphor and Thought*, 92–123. Cambridge, England: Cambridge University Press.

Spanier, E. 1966. *Algebraic Topology*. New York: McGraw Hill.

Sperber, D., and D. Wilson. 1986. *Relevance, Communication and Cognition*. Oxford: Basil Blackwell.

Tannen, D. 1979. *Processes and Consequences of Conversational Style*. PhD thesis, University of California, Berkeley, CA.

Turbayne, C. 1962. *The Myth of Metaphor*. Columbia S.C.: University of South Carolina Press.

Tversky, A. 1977. Features of Similarity. *Psychological Review* 84(4):327–352.

Urban, W. 1939. *Language and Reality*. London: G. Allen and Unwin, Ltd.

Vico, G. 1744. *The New Science of Giambattista Vico*. Ithaca NY: Cornell University Press, 1968. Translated by T. Bergin and M. Frisch.

Walton, K. 1990. *Mimesis as Make-Believe: On the Foundation of the Representational Arts.* Cambridge, MA: Harvard University.

Weld, D. S., and J. de Kleer (Eds.). 1989. *Readings in Qualitative Reasoning about Physical Systems.* San Mateo, CA: Morgan Kaufman Publishers, Inc.

Whorf, B. 1939. The Relation of Habitual Thought and Behavior to Language. In J. Carroll (Ed.), *Language, Thought, and Reality: Selected Writings of Benjamin Lee Whorf,* 134–159. New York: John Wiley and Sons. 1956.

Wilensky, R. 1983. *Planning and Understanding: A Computational Approach to Human Reasoning.* Reading, MA: Addison-Wesley.

Winston, P. 1978. Learning by Creating and Justifying Transfer Frames. *Artificial Intelligence* 10:147–172.

Woodhouse, A. S. P., and D. Bush (Eds.). 1972. *The Minor English Poems.* Vol. 2, Pt. 2 of *A Variorum Commentary on the Poems of John Milton.* New York: Columbia University Press.

Woods, W. 1970. Transition Network Grammars for Natural Language Analysis. *Communications of the ACM* 13:591–606.

CSLI Publications

Reports

The following titles have been published in the CSLI Reports series. These reports may be obtained from CSLI Publications, Ventura Hall, Stanford University, Stanford, CA 94305-4115.

The Situation in Logic–I Jon Barwise CSLI–84–2 (*$2.00*)

Coordination and How to Distinguish Categories Ivan Sag, Gerald Gazdar, Thomas Wasow, and Steven Weisler CSLI–84–3 (*$3.50*)

Belief and Incompleteness Kurt Konolige CSLI–84–4 (*$4.50*)

Equality, Types, Modules and Generics for Logic Programming Joseph Goguen and José Meseguer CSLI–84–5 (*$2.50*)

Lessons from Bolzano Johan van Benthem CSLI–84–6 (*$1.50*)

Self-propagating Search: A Unified Theory of Memory Pentti Kanerva CSLI–84–7 (*$9.00*)

Reflection and Semantics in LISP Brian Cantwell Smith CSLI–84–8 (*$2.50*)

The Implementation of Procedurally Reflective Languages Jim des Rivières and Brian Cantwell Smith CSLI–84–9 (*$3.00*)

Parameterized Programming Joseph Goguen CSLI–84–10 (*$3.50*)

Shifting Situations and Shaken Attitudes Jon Barwise and John Perry CSLI–84–13 (*$4.50*)

Completeness of Many-Sorted Equational Logic Joseph Goguen and José Meseguer CSLI–84–15 (*$2.50*)

Moving the Semantic Fulcrum Terry Winograd CSLI–84–17 (*$1.50*)

On the Mathematical Properties of Linguistic Theories C. Raymond Perrault CSLI–84–18 (*$3.00*)

A Simple and Efficient Implementation of Higher-order Functions in LISP Michael P. Georgeff and Stephen F.Bodnar CSLI–84–19 (*$4.50*)

On the Axiomatization of "if-then-else" Irène Guessarian and José Meseguer CSLI–85–20 (*$3.00*)

The Situation in Logic–II: Conditionals and Conditional Information Jon Barwise CSLI–84–21 (*$3.00*)

Principles of OBJ2 Kokichi Futatsugi, Joseph A. Goguen, Jean-Pierre Jouannaud, and José Meseguer CSLI–85–22 (*$2.00*)

Querying Logical Databases Moshe Vardi CSLI–85–23 (*$1.50*)

Computationally Relevant Properties of Natural Languages and Their Grammar Gerald Gazdar and Geoff Pullum CSLI–85–24 (*$3.50*)

An Internal Semantics for Modal Logic: Preliminary Report Ronald Fagin and Moshe Vardi CSLI–85–25 (*$2.00*)

The Situation in Logic–III: Situations, Sets and the Axiom of Foundation Jon Barwise CSLI–85–26 (*$2.50*)

Semantic Automata Johan van Benthem CSLI–85–27 (*$2.50*)

Restrictive and Non-Restrictive Modification Peter Sells CSLI–85–28 (*$3.00*)

Institutions: Abstract Model Theory for Computer Science J. A. Goguen and R. M. Burstall CSLI–85–30 (*$4.50*)

A Formal Theory of Knowledge and Action Robert C. Moore CSLI–85–31 (*$5.50*)

Finite State Morphology: A Review of Koskenniemi (1983) Gerald Gazdar CSLI–85–32 (*$1.50*)

The Role of Logic in Artificial Intelligence Robert C. Moore CSLI–85–33 (*$2.00*)

Lecture Notes

The titles in this series are distributed by the University of Chicago Press and may be purchased in academic or university bookstores or ordered directly from the distributor at 5801 Ellis Avenue, Chicago, Illinois 60637.

An Introduction to Unification-Based Approaches to Grammar Stuart M. Shieber. Lecture Notes No. 4

The Semantics of Destructive Lisp Ian A. Mason. Lecture Notes No. 5

An Essay on Facts Ken Olson. Lecture Notes No. 6

Logics of Time and Computation Robert Goldblatt. Lecture Notes No. 7

Word Order and Constituent Structure in German Hans Uszkoreit. Lecture Notes No. 8

Color and Color Perception: A Study in Anthropocentric Realism David Russel Hilbert. Lecture Notes No. 9

Prolog and Natural-Language Analysis Fernando C. N. Pereira and Stuart M. Shieber. Lecture Notes No. 10

Working Papers in Grammatical Theory and Discourse Structure: Interactions of Morphology, Syntax, and Discourse M. Iida, S. Wechsler, and D. Zec (Eds.) with an Introduction by Joan Bresnan. Lecture Notes No. 11

Natural Language Processing in the 1980s: A Bibliography Gerald Gazdar, Alex Franz, Karen Osborne, and Roger Evans. Lecture Notes No. 12

Information-Based Syntax and Semantics Carl Pollard and Ivan Sag. Lecture Notes No. 13

Non-Well-Founded Sets Peter Aczel. Lecture Notes No. 14

Partiality, Truth and Persistence Tore Langholm. Lecture Notes No. 15

Attribute-Value Logic and the Theory of Grammar Mark Johnson. Lecture Notes No. 16

The Situation in Logic Jon Barwise. Lecture Notes No. 17

The Linguistics of Punctuation Geoff Nunberg. Lecture Notes No. 18

Anaphora and Quantification in Situation Semantics Jean Mark Gawron and Stanley Peters. Lecture Notes No. 19

Propositional Attitudes: The Role of Content in Logic, Language, and Mind C. Anthony Anderson and Joseph Owens. Lecture Notes No. 20

Literature and Cognition Jerry R. Hobbs. Lecture Notes No. 21

Other CSLI Titles Distributed by UCP

Agreement in Natural Language: Approaches, Theories, Descriptions Michael Barlow and Charles A. Ferguson (Eds.)

Papers from the Second International Workshop on Japanese Syntax William J. Poser (Ed.)

The Proceedings of the Seventh West Coast Conference on Formal Linguistics (WCCFL 7)

The Proceedings of the Eighth West Coast Conference on Formal Linguistics (WCCFL 8)

The Phonology-Syntax Connection Sharon Inkelas and Draga Zec (Eds.) (co-published with The University of Chicago Press)

Books Distributed by CSLI

Titles distributed by CSLI may be ordered directly from CSLI Publications, Ventura Hall, Stanford University, Stanford, California 94305-4115.

The Proceedings of the Third West Coast Conference on Formal Linguistics (WCCFL 3) ($9.00)

The Proceedings of the Fourth West Coast Conference on Formal Linguistics (WCCFL 4) ($10.00)

The Proceedings of the Fifth West Coast Conference on Formal Linguistics (WCCFL 5) ($9.00)

The Proceedings of the Sixth West Coast Conference on Formal Linguistics (WCCFL 6) ($12.00)